50

mathematical
ideas

you really need to know

Tony Crilly

Quercus

Contents

Introduction

Mathematics is a vast subject and no one can possibly know it all. What one can do is explore and find an individual pathway. The possibilities open to us here will lead to other times and different cultures and to ideas that have intrigued mathematicians for centuries.

Mathematics is both ancient and modern and is built up from widespread cultural and political influences. From India and Arabia we derive our modern numbering system but it is one tempered with historical barnacles. The 'base 60' of the Babylonians of two or three millennia BC shows up in our own culture – we have 60 seconds in a minute and 60 minutes in an hour; a right angle is still 90 degrees and not 100 grads as revolutionary France adopted in a first move towards decimalization.

The technological triumphs of the modern age depend on mathematics and surely there is no longer any pride left in announcing to have been no good at it when at school. Of course school mathematics is a different thing, often taught with an eye to examinations. The time pressure of school does not help either, for mathematics is a subject where there is no merit in being fast. People need time to allow the ideas to sink in. Some of the greatest mathematicians have been painfully slow as they strove to understand the deep concepts of their subject.

There is no hurry with this book. It can be dipped into at leisure. Take your time and discover what these ideas you may have heard of really mean. Beginning with Zero, or elsewhere if you wish, you can move on a trip between islands of mathematical ideas. For instance, you can become knowledgeable about Game theory and next read about Magic squares. Alternatively you can move from Golden rectangles to the famous Fermat's last theorem, or any other path.

This is an exciting time for mathematics. Some of its major problems have been solved in recent times. Modern computing developments have helped with some but been helpless against others. The Four-colour problem was solved with the aid of a computer, but the Riemann hypothesis, the final chapter of the book, remains unsolved – by computer or any other means.

Mathematics is for all. The popularity of Sudoku is evidence that people can do mathematics (without knowing it) and enjoy it too. In mathematics, like art or music, there have been the geniuses but theirs is not the whole story. You will see several leaders making entrances and exits in some chapters only to reappear in others. Leonhard Euler, whose tercentenary occurs in 2007, is a frequent visitor to these pages. But, real progress in mathematics is the work of 'the many' accumulated over centuries. The choice of 50 topics is a personal one but I have tried to keep a balance. There are everyday and advanced items, pure and applied mathematics, abstract and concrete, the old and the new. Mathematics though is one united subject and the difficulty in writing has not been in choosing topics, but in leaving some out. There could have been 500 ideas but 50 are enough for a good beginning to your mathematical career.

01 Zero

At a young age we make an unsteady entrance into numberland. We learn that 1 is first in the 'number alphabet', and that it introduces the counting numbers 1, 2, 3, 4, 5, … Counting numbers are just that: they count real things – apples, oranges, bananas, pears. It is only later that we can count the number of apples in a box when there are none.

Even the early Greeks, who advanced science and mathematics by quantum leaps, and the Romans, renowned for their feats of engineering, lacked an effective way of dealing with the number of apples in an empty box. They failed to give 'nothing' a name. The Romans had their ways of combining I, V, X, L, C, D and M but where was 0? They did not count 'nothing'.

How did zero become accepted? The use of a symbol designating 'nothingness' is thought to have originated thousands of years ago. The Maya civilization in what is now Mexico used zero in various forms. A little later, the astronomer Claudius Ptolemy, influenced by the Babylonians, used a symbol akin to our modern 0 as a placeholder in his number system. As a placeholder, zero could be used to distinguish between examples (in modern notation) such as 75 and 705, instead of relying on context as the Babylonians had done. This might be compared with the introduction of the 'comma' into language – both help with *reading* the right meaning. But, just as the comma comes with a set of rules for its use – there have to be rules for using zero.

The seventh-century Indian mathematician Brahmagupta treated zero as a 'number', not merely as a placeholder, and set out rules for dealing with it. These included 'the sum of a positive number and zero is positive' and 'the sum of zero and zero is zero'. In thinking of zero as a number rather than a placeholder, he was quite advanced. The Hindu-Arabic numbering system which included zero in this way was promulgated in the West by Leonardo of Pisa – Fibonacci – in his *Liber Abaci* (*The Book of Counting*) first published in 1202. Brought up in North Africa and schooled in the Hindu-Arabian

timeline

700BC

The Babylonians use zero as a placeholder in their number system

AD628

Brahmagupta uses zero and states rules for its use with other numerals arithmetic, he recognized the power of using the extra sign 0 combined with the Hindu symbols 1, 2, 3, 4, 5, 6, 7, 8 and 9.

The launch of zero into the number system posed a problem which Brahmagupta had briefly addressed: how was this 'interloper' to be treated? He had made a start but his nostrums were vague. How could zero be integrated into the existing system of arithmetic in a more precise way? Some adjustments were straightforward. When it came to addition and multiplication, 0 fitted in neatly, but the operations of subtraction and division did not sit easily with the 'foreigner'. Meanings were needed to ensure that 0 harmonized with the rest of accepted arithmetic.

How does zero work? Adding and multiplying with zero is straightforward and uncontentious – you can add 0 to 10 to get a hundred – but we shall mean 'add' in the less imaginative way of the numerical operation. Adding 0 to a number leaves that number unchanged while multiplying 0 by any number always gives 0 as the answer. For example, we have $7 + 0 = 7$ and $7 \times 0 = 0$. Subtraction is a simple operation but can lead to negatives, $7 - 0 = 7$ and $0 - 7 = -7$, while division involving zero raises difficulties.

Let's imagine a length to be measured with a measuring rod. Suppose the measuring rod is actually 7 units in length. We are interested in how many measuring rods we can lie along our given length. If the length to be measured is actually 28 units the answer is 28 divided by 7 or in symbols $28 \div 7 = 4$. A better notation to express this division is

$$\frac{28}{7} = 4$$

and then we can 'cross-multiply' to write this in terms of multiplication, as $28 = 7 \times 4$. What now can be made of 0 divided by 7? To help suggest an answer in this case let us call the answer a so that

$$\frac{0}{7} = a$$

By cross-multiplication this is equivalent to $0 = 7 \times a$. If this is the case, the

830

Mahavira has ideas on how zero interacts with other numerals

1100

Bhaskara uses 0 as a symbol in algebra and attempts to show how it is manipulated

1202

Fibonacci uses the extra symbol 0 added to the Hindu-Arabic system of numerals 1,...,9 but not as a number on a par with them

only possible value for *a* is 0 itself because if the multiplication of two numbers gives 0, one of them must be 0. Clearly it is not 7 so *a* must be a zero.

This is *not* the main difficulty with zero. The danger point is division *by* 0. If we attempt to treat $\frac{7}{0}$ in the same way as we did with $\frac{0}{7}$, we would have the equation

$$\frac{7}{0} = b$$

By cross-multiplication, $0 \times b = 7$ and we wind up with the nonsense that $0 = 7$. By admitting the possibility of $\frac{7}{0}$ being a number we have the potential for numerical mayhem on a grand scale. The way out of this is to say that $\frac{7}{0}$ is undefined. It is not permissible to get any sense from the operation of dividing 7 (or any other nonzero number) by 0 and so we simply do not allow this operation to take place. In a similar way it is not permissible to place a comma in the mid,dle of a word without descending into nonsense.

The 12th-century Indian mathematician Bhaskara, following in the footsteps of Brahmagupta, considered division by 0 and suggested that a number divided by 0 was infinite. This is reasonable because if we divide a number by a very small number the answer is very large. For example, 7 divided by a tenth is 70, and by a hundredth is 700. By making the denominator number smaller and smaller the answer we get is larger and larger. In the ultimate smallness, 0 itself, the answer should be infinity. By adopting this form of reasoning, we are put in the position of explaining an even more bizarre concept – that is, infinity. Wrestling with infinity does not help; infinity (with its standard notation ∞) does not conform to the usual rules of arithmetic and is not a number in the usual sense.

If $\frac{7}{0}$ presented a problem, what can be done with the even more bizarre $\frac{0}{0}$? If $\frac{0}{0} = c$, by cross-multiplication, we arrive at the equation $0 = 0 \times c$ and the fact that $0 = 0$. This is not particularly illuminating but it is not nonsense either. In fact, *c* can be *any number* and we do not arrive at an impossibility. We reach the conclusion that $\frac{0}{0}$ can be anything; in polite mathematical circles it is called 'indeterminate'.

All in all, when we consider dividing by zero we arrive at the conclusion that it is best to exclude the operation from the way we do calculations. Arithmetic can be conducted quite happily without it.

What use is zero? We simply could not do without 0. The progress of science has depended on it. We talk about zero degrees longitude, zero degrees on the temperature scale, and likewise zero energy, and zero gravity. It has entered the non-scientific language with such ideas as the zero-hour and zero-tolerance.

Greater use could be made of it though. If you step off the 5th Ave sidewalk in New York City and into the Empire State Building, you are in the magnificent entrance lobby on Floor Number 1. This makes use of the ability of numbers to order, 1 for 'first', 2 for 'second' and so on, up to 102 for 'a hundred and second.' In Europe they do have a Floor 0 but there is a reluctance to call it that.

Mathematics could not function without zero. It is in the kernel of mathematical concepts which make the number system, algebra, and geometry go round. On the number line 0 is the number that separates the positive numbers from the negatives and thus occupies a privileged position. In the decimal system, zero serves as a place holder which enables us to use both huge numbers and microscopic figures.

> ## All about nothing
>
> The sum of zero and a positive number is positive
>
> The sum of zero and a negative number is negative
>
> The sum of a positive and a negative is their difference; or, if they are equal, zero
>
> Zero divided by a negative or positive number is either zero or is expressed as a fraction with zero as numerator and the finite quantity as denominator
>
> ## Brahmagupta, AD628

Over the course of hundreds of years zero has become accepted and utilized, becoming one of the greatest inventions of man. The 19th-century American mathematician G.B. Halsted adapted Shakespeare's *Midsummer Night's Dream* to write of it as the engine of progress that gives 'to airy nothing, not merely a local habitation and a name, a picture, a symbol, but helpful power, is the characteristic of the Hindu race from whence it sprang'.

When 0 was introduced it must have been thought odd, but mathematicians have a habit of fastening onto strange concepts which are proved useful much later. The modern day equivalent occurs in set theory where the concept of a *set* is a collection of elements. In this theory ϕ designates the set without any elements at all, the so-called 'empty set'. Now that is an odd idea, but like 0 it is indispensible.

the condensed idea
Nothing is quite something

02 Number systems

A number system is a method for handling the concept of 'how many'. Different cultures at differing periods of time have adopted various methods, ranging from the basic 'one, two, three, many' to the highly sophisticated decimal positional notation we use today.

The Sumerians and Babylonians, who inhabited present-day Syria, Jordan and Iraq around 4000 years ago, used a place-value system for their practical everyday use. We call it a place-value system because you can tell the 'number' by the positioning of a symbol. They also used 60 as the basic unit – what we call today a 'base 60' system. Vestiges of base 60 are still with us: 60 seconds in a minute, 60 minutes in an hour. When measuring angles we still reckon the full angle to be 360 degrees, despite the attempt of the metric system to make it 400 grads (so that each right angle is equal to 100 grads).

While our ancient ancestors primarily wanted numbers for practical ends, there is some evidence that these early cultures were intrigued by mathematics itself, and they took time off from the practicalities of life to explore them. These explorations included what we might call 'algebra' and also the properties of geometrical figures.

The Egyptian system from the 13th century BC used base ten with a system of hieroglyphic signs. Notably the Egyptians developed a system for dealing with fractions, but today's place-value decimal notation came from the Babylonians, later refined by the Hindus. Where it has the advantage is the way it can be used to express both very small and very large numbers. Using only the Hindu-Arabic numerals 1, 2, 3, 4, 5, 6, 7, 8 and 9, computations can be made with relative ease. To see this let's look at the Roman system. It suited their needs but only specialists in the system were capable of performing calculations with it.

timeline

30,000BC	2000BC
Palaeolithic peoples in Europe make number marks on bones	The Babylonians use symbols for numbers

The Roman system The basic symbols used by the Romans were the 'tens' (*I*, *X*, *C* and *M*), and the 'halves' of these (*V*, *L* and *D*). The symbols are combined to form others. It has been suggested that the use of *I*, *II*, *III* and *IIII* derives from the appearance of our fingers, *V* from the shape of the hand, and by inverting it and joining the two together to form the *X* we get two hands or ten fingers. *C* comes from *centum* and *M* from *mille*, the Latin for one hundred and one thousand respectively. The Romans also used *S* for 'a half' and a system of fractions based on 12.

The Roman system made some use of a 'before and after' method of producing the symbols needed but it would seem this was not uniformly adopted. The ancient Romans preferred to write *IIII* with *IV* only being introduced later. The combination *IX* seems to have been used, but a Roman would mean 8½ if *SIX* were written! Here are the *basic* numbers of the Roman system, with some additions from medieval times:

It's not easy handling Roman numerals. For example, the meaning of *MMMCDXLIIII* only becomes transparent when brackets are mentally introduced so that (*MMM*)(*CD*)(*XL*)(*IIII*) is then read as 3000 + 400 + 40 + 4 = 3444. But try adding *MMMCDXLIIII* + *CCCXCIIII*. A Roman skilled in the art would have short cuts and tricks, but for us it's difficult to obtain the right answer without first calculating it in the decimal system and translating the result back into Roman notation:

Roman number system

Roman Empire	medieval appendages
S a half	
I one	
V five	$\bar{V}$ five thousand
X ten	$\bar{X}$ ten thousand
L fifty	$\bar{L}$ fifty thousand
C hundred	$\bar{C}$ hundred thousand
D five hundred	$\bar{D}$ five hundred thousand
M thousand	$\bar{M}$ one million

Addition

3444	→	*MMMCDXLIIII*
+ 394	→	*CCCXCIIII*
=3838	→	*MMMDCCCXXXVIII*

The multiplication of two numbers is much more difficult and might be impossible within the basic system, even to Romans! To multiply 3444 × 394 we need the medieval appendages.

The forerunner of our modern decimal notation is used in India

The Hindu-Arabic system of writing numerals 1,…, 9, and a zero, spreads

The symbols of the decimal system take their recognizable modern forms

Multiplication

3444	→	MMMCDXLIIII
× 394	→	CCCXCIIII
=1,356,936	→	$\overline{MCCCLV}MCMXXXVI$

The Romans had no specific symbol for zero. If you asked a vegetarian citizen of Rome to record how many bottles of wine he'd consumed that day, he might write III but if you asked him how many chickens he'd eaten, he couldn't write 0. Vestiges of the Roman system survive in the pagination of some books (though not this one) and on the foundation stones of buildings. Some constructions were never used by the Romans, like MCM for 1900, but were introduced for stylistic reasons in modern times. The Romans would have written MDCCCC. The fourteenth King Louis of France, now universally known as Louis XIV, actually preferred to be known as Louis XIIII and made it a rule that his clocks were to show 4 o'clock as IIII o'clock.

A Louis XIIII clock

Decimal whole numbers
We naturally identify 'numbers' with decimal numbers. The decimal system is based on ten using the numerals 0, 1, 2, 3, 4, 5, 6, 7, 8 and 9. Actually it is based on 'tens' and 'units' but units can be absorbed into 'base 10'. When we write down the number **394**, we can explain its decimal meaning by saying it is composed of **3** hundreds, **9** tens and **4** units, and we could write

$$394 = 3 \times 100 + 9 \times 10 + 4 \times 1$$

This can be written using 'powers' of 10 (also known as 'exponentials' or 'indices'),

$$394 = 3 \times 10^2 + 9 \times 10^1 + 4 \times 10^0$$

where $10^2 = 10 \times 10$, $10^1 = 10$ and we agree separately that $10^0 = 1$. In this expression we see more clearly the *decimal* basis for our everyday number system, a system which makes addition and multiplication fairly transparent.

The point of decimal
So far we have looked at representing whole numbers. Can the decimal system cope with parts of a number, like $^{572}/_{1000}$? This means

$$\frac{572}{1000} = \frac{5}{10} + \frac{7}{100} + \frac{2}{1000}$$

We can treat the 'reciprocals' of 10, 100, 1000 as *negative* powers of 10, so that

$$\frac{572}{1000} = 5 \times 10^{-1} + 7 \times 10^{-2} + 2 \times 10^{-3}$$

and this can be written **.572** where the decimal point indicates the beginning of the negative powers of 10. If we add this to the decimal expression for 394 we get the decimal expansion for the number 394$^{572}/_{1000}$, which is simply **394.572**.

For very big numbers the decimal notation can be very long, so we revert in this case to the 'scientific notation'. For example, 1,356,936,892 can be written as 1.356936892×10^9 which often appears as '1.356936892 × 10E9' on calculators or computers. Here, the power 9 is one less than the number of digits in the number and the letter E stands for 'exponential'. Sometimes we might want to use bigger numbers still, for instance if we were talking about the number of hydrogen atoms in the known universe. This has been estimated as about 1.7×10^{77}. Equally 1.7×10^{-77}, with a negative power, is a very small number and this too is easily handled using scientific notation. We couldn't begin to think of these numbers with the Roman symbols.

Zeros and ones While base 10 is common currency in everyday life, some applications require other bases. The binary system which uses base 2 lies behind the power of the modern computer. The beauty of binary is that any number can be expressed using only the symbols 0 and 1. The tradeoff for this economy is that the number expressions can be very long.

How can we express **394** in binary notation? This time we are dealing with powers of 2 and after some working out we can give the full expression as,

$$394 = 1 \times 256 + 1 \times 128 + 0 \times 64 + 0 \times 32 + 0 \times 16 + 1 \times 8 + 0 \times 4 + 1 \times 2 + 0 \times 1$$

so that reading off the zeros and ones, **394** in binary is **110001010**.

Powers of 2	Decimal
2^0	1
2^1	2
2^2	4
2^3	8
2^4	16
2^5	32
2^6	64
2^7	128
2^8	256
2^9	512
2^{10}	1024

As binary expressions can be very long, other bases frequently arise in computing. These are the octal system (base 8) and the hexadecimal system (base 16). In the octal system we only need the symbols 0, 1, 2, 3, 4, 5, 6, 7, whereas hexadecimal uses 16 symbols. In this base 16 system, we customarily use 0, 1, 2, 3, 4, 5, 6, 7, 8, 9, A, B, C, D, E, F. As 10 corresponds to the letter A, the number **394** is represented in hexadecimal as 18A. It's as easy as ABC, which bear in mind, is really 2748 in decimal!

the condensed idea
Writing numbers down

03 Fractions

A fraction is a 'fractured number' – literally. If we break up a whole number an appropriate way to do it is to use fractions. Let's take the traditional example, the celebrated cake, and break it into three parts.

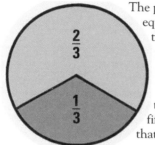

The person who gets two of the three parts of the cake gets a fraction equivalent to ⅔. The unlucky person only gets ⅓. Putting together the two portions of the cake we get the whole cake back, or in fractions, ⅓ + ⅔ = 1 where 1 represents the whole cake.

Here is another example. You might have been to the sales and seen a shirt advertised at four-fifths of the original price. Here the fraction is written as ⅘. We could also say the shirt has a fifth off the original price. That would be written as ⅕ and we see that ⅕ + ⅘ = 1 where 1 represents the original price.

A fraction always has the form of a whole number 'over' a whole number. The bottom number is called the 'denominator' because it tells us how many parts make the whole. The top number is called the 'numerator' because it tells us how many unit fractions there are. So a fraction in established notation always looks like

$$\frac{\text{numerator}}{\text{denominator}}$$

In the case of the cake, the fraction you might want to eat is ⅔ where the denominator is 3 and the numerator is 2. ⅔ is made up of 2 unit fractions of ⅓.

We can also have fractions like ¹⁴⁄₅ (called improper fractions) where the numerator is bigger than the denominator. Dividing 14 by 5 we get 2 with 4 left over, which can be written as the 'mixed' number 2⅘. This comprises the whole number 2 and the 'proper' fraction ⅘. Some early writers wrote this as ⅘2. Fractions are usually represented in a form where the numerator and denominator (the 'top' and the 'bottom') have no common factors. For

timeline

1800 BC	1650 BC
Fractions are used in Babylonian cultures	The Egyptians make use of unit fractions

example, the numerator and denominator of $\frac{8}{10}$ have a common factor of 2, because $8 = 2 \times 4$ and $10 = 2 \times 5$. If we write the fraction $\frac{8}{10} = \frac{2 \times 4}{2 \times 5}$ we can 'cancel' the 2s out and so $\frac{8}{10} = \frac{4}{5}$, a simpler form with the same value. Mathematicians refer to fractions as 'rational numbers' because they are ratios of two numbers. The rational numbers were the numbers the Greeks could 'measure'.

Adding and multiplying The rather curious thing about fractions is that they are easier to multiply than to add. Multiplication of whole numbers is so troublesome that ingenious ways had to be invented to do it. But with fractions, it's addition that's more difficult and takes some thinking about.

Let's start by multiplying fractions. If you buy a shirt at four-fifths of the original price of £30 you end up paying the sale price of £24. The £30 is divided into five parts of £6 each and four of these five parts is $4 \times 6 = 24$, the amount you pay for the shirt.

Subsequently, the manager of the shop discovers that the shirts are not selling at all well so he drops the price still further, advertising them at $\frac{1}{2}$ of the sale price. If you go into the shop you can now get the shirt for £12. This is $\frac{1}{2} \times \frac{4}{5} \times 30$ which is equal to 12. To multiply two fractions together you just multiply the denominators together and the numerators together:

$$\frac{1}{2} \times \frac{4}{5} = \frac{1 \times 4}{2 \times 5} = \frac{4}{10}$$

If the manager had made the two reductions at a single stroke he would have advertised the shirts at four-tenths of the original price of £30. This is $\frac{4}{10} \times 30$ which is £12.

Adding two fractions is a different proposition. The addition $\frac{1}{3} + \frac{2}{3}$ is OK because the denominators are the same. We simply add the two numerators together to get $\frac{3}{3}$, or 1. But how could we add two-thirds of a cake to four-fifths of a cake? How could we figure out $\frac{2}{3} + \frac{4}{5}$? If only we could say $\frac{2}{3} + \frac{4}{5} = \frac{2+4}{3+5} = \frac{6}{8}$ but unfortunately we cannot.

Adding fractions requires a different approach. To add $\frac{2}{3}$ and $\frac{4}{5}$ we must first express each of them as fractions which have the *same* denominators. First multiply the top and bottom of ⅔ by 5 to get ¹⁰⁄₁₅. Now multiply the top and bottom of ⅘ by 3 to get ¹²⁄₁₅. Now both fractions have 15 as a common denominator and to add them we just add the new numerators together:

$$\frac{2}{3} + \frac{4}{5} = \frac{10}{15} + \frac{12}{15} = \frac{22}{15}$$

Converting to decimals In the world of science and most applications of mathematics, decimals are the preferred way of expressing fractions. The fraction ⅘ is the same as the fraction ⁸⁄₁₀ which has 10 as a denominator and we can write this as the decimal 0.8.

Fractions which have 5 or 10 as a denominator are easy to convert. But how could we convert, say ⅞, into decimal form? All we need to know is that when we divide a whole number by another, either it goes in exactly or it goes in a certain number of times with something left over, which we call the 'remainder'.

Using ⅞ as our example, the recipe to convert from fractions to decimals goes like this:

• Try to divide 8 into 7. It doesn't go, or you could say it goes 0 times with remainder 7. We record this by writing zero followed by the decimal point: '0.'
• Now divide 8 into 70 (the remainder of the previous step multiplied by 10). This goes 8 times, since 8 × 8 = 64, so the answer is 8 with remainder 6 (70 − 64). So we write this alongside our first step, to make '0.8'
• Now divide 8 into 60 (the remainder of the previous step multiplied by 10). Because 7 × 8 = 56, the answer is 7 with remainder 4. We write this down, and so far we have '0.87'
• Divide 8 into 40 (the remainder of the previous step multiplied by 10). The answer is *exactly* 5 with remainder *zero*. When we get remainder 0 the recipe is complete. We are finished. The final answer is '0.875'.

When applying this conversion recipe to other fractions it is possible that we might never finish! We could keep going forever; if we try to convert ⅔ into decimal, for instance, we find that at each stage the result of dividing 20 by 3 is 6 with a remainder of 2. So we have again to divide 6 into 20, and we never get to the point where the remainder is 0. In this case we have the infinite decimal 0.666666… This is written 0.6̇ to indicate the 'recurring decimal'.

There are many fractions that lead us on forever like this. The fraction $\frac{5}{7}$ is interesting. In this case we get $\frac{5}{7}=0.714285714285714285...$ and we see that the sequence 714285 keeps repeating itself. If *any* fraction results in a repeating sequence we cannot ever write it down in a terminating decimal and the 'dotty' notation comes into its own. In the case of $\frac{5}{7}$ we write $\frac{5}{7} = .\overline{714285}$.

Egyptian fractions The Egyptians of the second millennium BC based their system of fractions on hieroglyphs designating *unit* fractions – those fractions whose numerators are 1. We know this from the Rhind Papyrus which is kept in the British Museum. It was such a complicated system of fractions that only those trained in its use could know its inner secrets and make the correct calculations.

The Egyptians used a few privileged fractions such as $\frac{2}{3}$ but all other fractions were expressed in terms of unit fractions like $\frac{1}{2}$, $\frac{1}{3}$, $\frac{1}{11}$ or $\frac{1}{168}$. These were their 'basic fractions' from which all other fractions could be expressed. For example $\frac{5}{7}$ is not a unit fraction but it could be written in terms of the unit fractions,

$$\tfrac{5}{7} = \tfrac{1}{3} + \tfrac{1}{4} + \tfrac{1}{8} + \tfrac{1}{168}$$

where *different* unit fractions must be used. A feature of the system is that there may be more than one way of writing a fraction, and some ways are shorter than others. For example,

$$\tfrac{5}{7} = \tfrac{1}{2} + \tfrac{1}{7} + \tfrac{1}{14}$$

The 'Egyptian expansion' may have had limited practical use but the system has inspired generations of pure mathematicians and provided many challenging problems, some of which remain unsolved today. For instance, a full analysis of the methods for finding the *shortest* Egyptian expansion awaits the intrepid mathematical explorer.

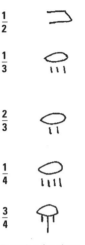

Egyptian fractions

the condensed idea
One number over another

04 Squares and square roots

If you like making squares with dots, your thought patterns are similar to those of the Pythagoreans. This activity was prized by the fraternity who followed their leader Pythagoras, a man best remembered for 'that theorem'. He was born on the Greek island of Samos and his secret religious society thrived in southern Italy. Pythagoreans believed mathematics was the key to the nature of the universe.

Counting up the dots, we see the first 'square' on the left is made from one dot. To the Pythagoreans 1 was the most important number, imbued with spiritual existence. So we've made a good start. Continuing to count up the dots of the subsequent squares gives us the 'square' numbers, 1, 4, 9, 16, 25, 36, 49, 64,…

These are called 'perfect' squares. You can compute a square number by adding the dots on the shape ⌐ outside the previous one, for example 9 + 7 = 16. The Pythagoreans didn't stop with squares. They considered other shapes, such as triangles, pentagons (the figure with five sides) and other polygonal (many-sided) shapes.

1   4   9   16

The triangular numbers resemble a pile of stones. Counting these dots gives us 1, 3, 6, 10, 15, 21, 28, 36, … If you want to compute a triangular number you can use the previous one and add the number of dots in the last row. What is the triangular number which comes after 10, for instance? It will have 5 dots in the last row so we just add 10 + 5 = 15.

1   3   6   10

If you compare the square and triangular numbers you will see that the number 36 appears in both lists. But there is a more striking link. If you take *successive*

timeline

1750BC	**525**BC	**c.300**BC
The Babylonians compile tables of square roots	The Pythagoreans study geometrically arranged square numbers	The theory of the irrational numbers by Eudoxus is published in Book 5 of Euclid's *Elements*

triangular numbers and add them together, what do you get? Let's try it and put the results in a table.

That's right! When you add two successive triangular numbers together you get a square number. You can also see this with a 'proof without words'. Consider a square made up of 4 rows of 4 dots with a diagonal line drawn through it. The dots above the line (shown) form a triangular number and below the line is the next triangular number. This observation holds for any sized square. It's a short step from these 'dotty diagrams' to measuring areas. The area of a square whose side is 4 is $4 \times 4 = 4^2 = 16$ square units. In general, if the side is called x then the area will be x^2.

The square x^2 is the basis for the parabolic shape. This is the shape you find in satellite receiver dishes or the reflector mirrors of car headlights. A parabola has a focus point. In a receiving dish a sensor placed at the focus receives the reflected signals when parallel beams from space hit the curved dish and bounce towards the focus point.

In a car headlight a light bulb at the focus *sends* out a parallel beam of light. In sport, shot-putters, javelin throwers and hammer throwers will all recognize the parabola as the shape of the curved path that every object follows as it falls to the Earth.

Square roots If we turn the question around and want to find the length of a square which has a given area 16, the answer is plainly 4. The square root of 16 is 4 and written as $\sqrt{16} = 4$. The symbol $\sqrt{\ }$ for square roots has been employed since the 1500s. All the square numbers have nice whole numbers as square roots. For example, $\sqrt{1} = 1$, $\sqrt{4} = 2$, $\sqrt{9} = 3$, $\sqrt{16} = 4$, $\sqrt{25} = 5$, and so on. There are though many gaps along the numbers line between these perfect squares. These are 2, 3, 5, 6, 7, 8, 10, 11, ...

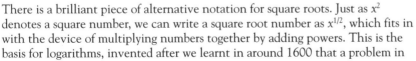

There is a brilliant piece of alternative notation for square roots. Just as x^2 denotes a square number, we can write a square root number as $x^{1/2}$, which fits in with the device of multiplying numbers together by adding powers. This is the basis for logarithms, invented after we learnt in around 1600 that a problem in

Adding two successive triangular numbers

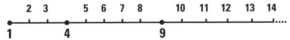

1 + 3	4
3 + 6	9
6 + 10	16
10 + 15	25
15 + 21	36
21 + 28	49
28 + 36	64

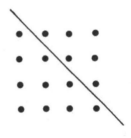

630

rahmagupta gives methods for
omputing square roots

1550

The symbol √ is introduced for
square roots

1872

Richard Dedekind sets out a
theory of irrational numbers

multiplication could be changed into one of addition. But that is another story. These numbers all have square roots, but they are not equal to whole numbers. Virtually all calculators have a √ button, and using it we find, for instance, √7 = 2.645751311.

Let's look at √2. The number 2 had special significance for the Pythagoreans because it is the first even number (the Greeks thought of the even numbers as feminine and the odd ones as masculine – and the small numbers had distinct personalities). If you work out √2 on your calculator you will get 1.414213562 assuming your calculator gives this many decimal places. Is this the square root of 2? To check we make the calculation 1.414213562 × 1.414213562. This turns out to be 1.999999999. This is not quite 2 for 1.414213562 is only an approximation for the square root of 2.

What is perhaps remarkable is that all we will ever get is an approximation! The decimal expansion of √2 to millions of decimal places will only ever be an approximation. The number √2 is important in mathematics, perhaps not quite as illustrious as π or e (see pages 20–27) but important enough to gets its own name – it is sometimes called the 'Pythagorean number'.

Are square roots fractions? Asking whether square roots are fractions is linked to the theory of measurement as known to the ancient Greeks. Suppose we have a line AB whose length we wish to measure, and an indivisible 'unit' CD with which to measure it. To make the measurement we

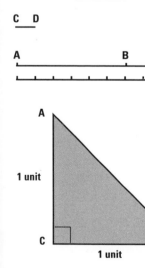

place the unit CD sequentially against AB. If we place the unit down m times *and* the end of the last unit fits flush with the end of AB (at the point B) then the length of AB will simply be m. If not we can place a copy of AB next to the original and carry on measuring with the unit (see figure). The Greeks believed that at some point using n copies of AB and m units, the unit would fit *flush* with the end-point of the mth AB. The length of AB would then be m/n. For example if 3 copies of AB are laid side by side and 29 units fit alongside, the length of AB would be 29/3.

The Greeks also considered how to measure the length of the side AB (the hypotenuse) of a triangle where both of the other sides are one 'unit' long. By Pythagoras's theorem the length of AB could be written symbolically as √2 so the question is whether √2 = m/n?

From our calculator, we have already seen that the decimal expression for √2 is potentially infinite, and this fact (that there is no end to the decimal expression) perhaps indicates that √2

fraction. But there is no end to the decimal 0.3333333… and that represents the fraction ⅓. We need more convincing arguments.

Is √2 a fraction? This brings us to one of the most famous proofs in mathematics. It follows the lines of the type of proof which the Greeks loved: the method of *reductio ad absurdum*. Firstly it is assumed that √2 cannot be a fraction and 'not a fraction' at the same time. This is the law of logic called the 'excluded middle'. There is no middle way in this logic. So what the Greeks did was ingenious. They assumed that it was a fraction and, by strict logic at every step, derived a contradiction, an 'absurdity'. So let's do it. Suppose

$$\sqrt{2} = \frac{m}{n}$$

We can assume a bit more too. We can assume that m and n have no common factors. This is OK because if they did have common factors these could be cancelled before we began. (For example, the fraction ²¹⁄₃₅ is equivalent to the factorless ⅗ on cancellation of the common factor 7.)

We can square both sides of √2 = $^m/_n$ to get 2 = $^{m^2}/_{n^2}$ and so $m^2 = 2n^2$. Here is where we make our first observation: since m^2 is 2 times something it must be an even number. Next m itself cannot be odd (because the square of an odd number is odd) and so m is also an even number.

So far the logic is impeccable. As m is even it must be twice something which we can write as $m = 2k$. Squaring both sides of this means that $m^2 = 4k^2$. Combining this with the fact that $m^2 = 2n^2$ means that $2n^2 = 4k^2$ and on cancellation of 2 we conclude that $n^2 = 2k^2$. But we have been here before! And as before we conclude that n^2 is even and n itself is even. We have thus deduced by strict logic that both m and n are both even and so they have a factor of 2 in common. This was contrary to our assumption that m and n have no common factors. The conclusion therefore is that √2 cannot be a fraction.

It can also be proved that the whole sequence of numbers √n (except where n is a perfect square) cannot be fractions. Numbers which cannot be expressed as fractions are called 'irrational' numbers, so we have observed there are an infinite number of irrational numbers.

the condensed idea
The way to irrational numbers

05 π

π is the most famous number in mathematics. Forget all the other constants of nature, π will always come at the top of the list. If there were Oscars for numbers, π would get an award every year.

π, or pi, is the length of the outside of a circle (the circumference) divided by the length across its centre (the diameter). Its value, the ratio of these two lengths, does not depend on the size of the circle. Whether the circle is big or small, π is indeed a mathematical constant. The circle is the natural habitat for π but it occurs everywhere in mathematics, and in places not remotely connected with the circle.

For a circle of diameter *d* and radius *r*:

$$\text{circumference} = \pi d = 2\pi r$$
$$\text{area} = \pi r^2$$

For a sphere of diameter *d* and radius *r*:

$$\text{surface area} = \pi d^2 = 4\pi r^2$$
$$\text{volume} = \frac{4}{3}\pi r^3$$

Archimedes of Syracuse The ratio of the circumference to the diameter of a circle was a subject of ancient interest. Around 2000 BC the Babylonians made the observation that the circumference was roughly 3 times as long as its diameter.

It was Archimedes of Syracuse who made a real start on the mathematical theory of π, in around 225 BC. Archimedes is right up there with the greats. Mathematicians love to rate their co-workers and they place him on a level with Carl Friedrich Gauss (The 'Prince of Mathematicians') and Sir Isaac Newton. Whatever the merits of this judgment it is clear that Archimedes would be in any mathematics Hall of Fame. He was hardly an ivory tower figure though – as well as his contributions to astronomy, mathematics and physics, he also designed weapons of war, such as catapults, levers and 'burning mirrors', all used to help keep the Romans at bay. But by all accounts he did have something of the absent-mindedness of the professor, for what else would induce him to leap from his bath and run naked down the street shouting 'Eureka' at discovering the law of buoyancy in hydrostatics? How he celebrated his work on π is not recorded.

timeline

2000 BC	250 BC
The Babylonians observe π is roughly 3	Archimedes gives the close approximation to π of 22/7

Given that *π* is defined as the ratio of its circumference to its diameter, what does it have to do with the *area* of a circle? It is a *deduction* that the area of a circle of radius r is πr^2, though this is probably better known than the circumference/diameter definition of *π*. The fact that *π* does double duty for area and circumference is remarkable.

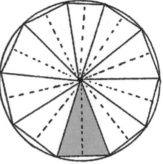

How can this be shown? The circle can be split up into a number of narrow equal triangles with base length *b* whose height is approximately the radius *r*. These form a polygon inside the circle which approximates the area of the circle. Let's take 1000 triangles for a start. The whole process is an exercise in approximations. We can join together each adjacent *pair* of these triangles to form a rectangle (approximately) with area $b \times r$ so that the total area of the polygon will be $500 \times b \times r$. As $500 \times b$ is about half the circumference it has length πr, the area of the polygon is $\pi r \times r = \pi r^2$. The more triangles we take the closer will be the approximation and in the limit we conclude the area of the circle is πr^2.

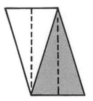

Archimedes estimated the value of *π* as bounded between $^{223}/_{71}$ and $^{220}/_{70}$. And so it is to Archimedes that we owe the familiar approximation 22/7 for the value of *π*. The honour for designating the actual symbol *π* goes to the little known William Jones, a Welsh mathematician who became Vice President of the Royal Society of London in the 18th century. It was the mathematician and physicist Leonhard Euler who popularized *π* in the context of the circle ratio.

The exact value of *π*   We can never know the *exact* value of *π* because it is an irrational number, a fact proved by Johann Lambert in 1768. The decimal expansion is infinite with no predictable pattern. The first 20 decimal places are 3.14159265358979323846…The value of √10 used by the Chinese mathematicians is 3.16227766016837933199 and this was adopted around AD 500 by Brahmagupta. This value is in fact little better than than the crude value of 3 and it differs in the second decimal place from *π*.

π can be computed from a *series* of numbers. A well known one is

$$\frac{\pi}{4} = 1 - \frac{1}{3} + \frac{1}{5} - \frac{1}{7} + \frac{1}{9} - \frac{1}{11} + \cdots$$

though this is painfully slow in its convergence on *π* and quite hopeless for

AD 1706

William Jones introduces the symbol *π*

1761

Lambert proves that *π* is irrational

1882

Lindemann proves that *π* is transcendental

calculation. Euler found a remarkable series that converges to π:

$$\frac{\pi^2}{6} = 1 + \frac{1}{2^2} + \frac{1}{3^2} + \frac{1}{4^2} + \frac{1}{5^2} + \frac{1}{6^2} + \cdots$$

The self-taught genius Srinivasa Ramanujan devised some spectacular approximating formulae for π. One involving only the square root of 2 is:

$$\frac{9801}{4412}\sqrt{2} = 3.1415927300133056603139961890\ldots$$

Mathematicians are fascinated by π. While Lambert had proved it could not be a fraction, in 1882 the German mathematician Ferdinand von Lindemann solved the most outstanding problem associated with π. He showed that π is 'transcendental'; that is, π cannot be the solution of an algebraic equation (an equation which only involves powers of x). By solving this 'riddle of the ages' Lindemann concluded the problem of 'squaring the circle'. Given a circle the challenge was to construct a square of the same area using only a pair of compasses and a straight edge. Lindemann proved conclusively that it cannot be done. Nowadays the phrase squaring the circle is the equivalent of an impossibility.

The actual calculation of π continued apace. In 1853, William Shanks claimed a value correct to 607 places (actually correct up to only 527). In modern times the quest for calculating π to more and more decimal places gained momentum through the modern computer. In 1949, π was calculated to 2037 decimal places, which took 70 hours to do on an ENIAC computer. By 2002, π had been computed to a staggering 1,241,100,000,000 places, but it is an ever growing tail. If we stood on the equator and started writing down the expansion of π, Shanks' calculation would take us a full 14 metres, but the length of the 2002 expansion would take us about 62 laps around the world!

Various questions about π have been asked and aswered. Are the digits of π random? Is it possible to find a predetermined sequence in the expansion? For instance, is it possible to find the sequence 0123456789 in the expansion? In the 1950s this seemed unknowable. No one had found such a sequence in the 2000 known digits of π. L.E.J. Brouwer, a leading Dutch mathematician, said the question was devoid of meaning since he believed it could not be experienced. In fact these digits were found in 1997 beginning at the position 17,387,594,880, or, using the equator metaphor, about 3000 miles before one lap is completed. You will find ten sixes in a row before you have completed 600 miles but will have to wait until one lap has been completed and gone a further 3600 miles to find ten sevens in a row.

π in poetry

If you really want to remember the first values in the expansion of π perhaps a little poetry will help. Following the tradition of teaching mathematics in the 'mnemonic way' there is a brilliant variation of Edgar Allen Poe's poem 'The Raven' by Michael Keith.

The real poem by Poe begins

The raven E.A. Poe

Once upon a midnight dreary, while I pondered weak and weary,

Over many a quaint and curious volume of forgotten lore,

Keith's variant for π begins

Poe, E. Near A Raven

Midnights so dreary, tired and weary.

Silently pondering volumes extolling all by-now obsolete lore.

The letter count of each successive word in Keith's version provides the first 740 digits of π.

The importance of π   What is the use of knowing π to so many places? After all, most calculations only require a few decimal places; probably no more than ten places are needed for any practical application, and Archimedes' approximation of 22/7 is good enough for most. But the extensive calculations are not just for fun. They are used to test the limits of computers, besides exerting a fascination on the group of mathematicians who have called themselves the 'friends of pi'.

Perhaps the strangest episode in the story of π was the attempt in the Indiana State Legislature to pass a bill that would fix its value. This occurred at the end of the 19th century when a medical doctor Dr E.J. Goodwin introduced the bill to make π 'digestible'. A practical problem encountered in this piece of legislation was the proposer's inability to fix the value he wanted. Happily for Indiana, the folly of legislating on π was realized before the bill was fully ratified. Since that day, politicians have left π well alone.

the condensed idea
When the π was opened

06 e

e is the new kid on the block when compared with its only rival π. While π is more august and has a grand past dating back to the Babylonians, e is not so weighed down by the barnacles of history. The constant e is youthful and vibrant and is ever present when 'growth' is involved. Whether it's populations, money or other physical quantities, growth invariably involves e.

e is the number whose approximate value is 2.71828. So why is that so special? It isn't a number picked out at random, but is one of the great mathematical constants. It came to light in the early 17th century when several mathematicians put their energies into clarifying the idea of a logarithm, the brilliant invention that allowed the multiplication of large numbers to be converted into addition.

But the story really begins with some 17th-century e-commerce. Jacob Bernoulli was one of the illustrious Bernoullis of Switzerland, a family which made it their business to supply a dynasty of mathematicians to the world. Jacob set to work in 1683 with the problem of compound interest.

Money, money, money Suppose we consider a 1-year time period, an interest rate of a whopping 100%, and an initial deposit (called a 'principal' sum) of £1. Of course we rarely get 100% on our money but this figure suits our purpose and the concept can be adapted to realistic interests rates like 6% and 7%. Likewise, if we have greater principal sums like £10,000 we can multiply everything we do by 10,000.

At the end of the year at 100% interest, we will have the principal and the amount of interest earned which in this case is also £1. So we shall have the princely sum of £2. Now we suppose that the interest rate is halved to 50% but is applied for each half-year separately. For the first half-year we gain an interest of 50 pence and our principal has grown to £1.50 by the end of the

timeline

first half-year. So, by the end of the full year we would have this amount and the 75 pence interest on this sum. Our £1 has grown to £2.25 by the end of the year! By compounding the interest each half-year we have made an extra 25 pence. It may not seem much but if we had £10,000 to invest, we would have £2,250 interest instead of £2,000. By compounding every half-year we gain an extra £250.

But if compounding every half-year means we gain on our savings, the bank will also gain on any money we owe – so we must be careful! Suppose now that the year is split into four quarters and 25% is applied to each quarter. Carrying out a similar calculation, we find that our £1 has grown to £2.44141. Our money is growing and with our £10,000 it would seem to be advantageous if we could split up the year and apply the smaller percentage interest rates to the smaller time intervals.

Will our money increase beyond all bounds and make us a millionaires? If we keep dividing the year up into smaller and smaller units, as shown in the table, this 'limiting process' shows that the amount appears to be settling down to a constant number. Of course, the only realistic compounding period is per day (and this is what banks do). The mathematical message is that this limit, which mathematicians call *e*, is the amount £1 grows to if compounding takes place continuously. Is this a good thing or a bad thing? You know the answer: if you are saving, 'yes'; if you owe money, 'no'. It's a matter of '*e*-learning'.

Compounding each …	Accrued sum
year	£2.00000
half-year	£2.25000
quarter	£2.44141
month	£2.61304
week	£2.69260
day	£2.71457
hour	£2.71813
minute	£2.71828
second	£2.71828

The exact value of *e* Like π, *e* is an irrational number so, as with π, we cannot know its exact value. To 20 decimal places, the value of *e* is 2.71828182845904523536…

Using only fractions, the best approximation to the value of *e* is 87/32 if the top and bottom of the fraction are limited to two-digit numbers. Curiously, if the top and bottom are limited to three-digit numbers the best fraction is 878/323. This second fraction is a sort of palindromic extension of the first one – mathematics has a habit of offering these little surprises. A well-known series expansion for *e* is given by

$$e = 1 + \frac{1}{1} + \frac{1}{2 \times 1} + \frac{1}{3 \times 2 \times 1} + \frac{1}{4 \times 3 \times 2 \times 1} + \frac{1}{5 \times 4 \times 3 \times 2 \times 1} \cdots$$

748

…ler calculates *e* to 23 digits; he is given the edit for the discovery of the famous formula ⁱ + 1 = 0 around this time

1873

Hermite proves *e* is a transcendental number

2007

e is calculated to the order of 10^{11} digits

The factorial notation using an exclamation mark is handy here. In this, for example, 5! = 5×4×3×2×1. Using this notation, e takes the more familiar form

$$e = 1 + \frac{1}{1!} + \frac{1}{2!} + \frac{1}{3!} + \frac{1}{4!} + \frac{1}{5!} + \cdots$$

So the number e certainly seems to have some pattern. In its mathematical properties, e appears more 'symmetric' than π.

If you want a way of remembering the first few places of e, try this: 'We attempt a mnemonic to remember a strategy to memorize this count…', where the letter count of each word gives the next number of e. If you know your American history then you might remember that e is '2.7 Andrew Jackson Andrew Jackson', because Andrew Jackson ('Old Hickory'), the seventh president of the United States was elected in 1828. There are many such devices for remembering e but their interest lies in their quaintness rather than any mathematical advantage.

That e is irrational (not a fraction) was proved by Leonhard Euler in 1737. In 1840, French mathematician Joseph Liouville showed that e was not the solution of any quadratic equation and in 1873, in a path-breaking work, his countryman Charles Hermite, proved that e is transcendental (it cannot be the solution of *any* algebraic equation). What was important here was the method Hermite used. Nine years later, Ferdinand von Lindemann adapted Hermites's method to prove that π was transcendental, a problem with a much higher profile.

One question was answered but new ones appeared. Is e raised to the power of e transcendental? It is such a bizarre expression, how could this be otherwise? Yet this has not been proved rigorously and, by the strict standards of mathematics, it must still be classified as a conjecture. Mathematicians have inched towards a proof, and have proved it is impossible for *both* it and e raised to the power of e^2 to be transcendental. Close, but not close enough.

The connections between π and e are fascinating. The values of e^π and π^e are close but it is easily shown (without actually calculating their values) that $e^\pi > \pi^e$. If you 'cheat' and have a look on your calculator, you will see that approximate values are $e^\pi = 23.14069$ and $\pi^e = 22.45916$.

The number e^π is known as Gelfond's constant (named after the Russian mathematician Aleksandr Gelfond) and has been shown to be a transcendental. Much less is known about π^e; it has not yet been proved to be irrational – if indeed it is.

Is *e* important? The chief place where *e* is found is in growth. Examples are economic growth and the growth of populations. Connected with this are the curves depending on *e* used to model radioactive decay.

The number *e* also occurs in problems not connected with growth. Pierre Montmort investigated a probability problem in the 18th century and it has since been studied extensively. In the simple version a group of people go to lunch and afterwards pick up their hats at random. What is the probability that no one gets their own hat?

It can be shown that this probability is ⅟ₑ (about 37%) so that the probability of at least one person getting their own hat is 1 – ⅟ₑ (63%). This application in probability theory is one of many. The Poisson distribution which deals with rare events is another. These were early instances but by no means isolated ones: James Stirling achieved a remarkable approximation to the factorial value n! involving *e* (and π); in statistics the familiar 'bell curve' of the normal distribution involves *e*; and in engineering the curve of a suspension bridge cable depends on *e*. The list is endless.

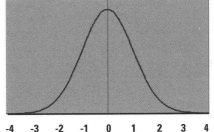

An earth-shattering identity The prize for the most remarkable formula of all mathematics involves *e*. When we think of the famous numbers of mathematics we think of 0, 1, π, *e* and the imaginary number $i = \sqrt{-1}$. How could it be that

$$e^{i\pi} + 1 = 0$$

-4 -3 -2 -1 0 1 2 3 4

The normal distribution

It is! This is a result attributed to Euler.

Perhaps *e*'s real importance lies in the mystery by which it has captivated generations of mathematicians. All in all, *e* is unavoidable. Just why an author like E.V. Wright should put himself through the effort of writing an e-less novel – presumably he had a pen name too – but his *Gadsby* is just that. It is hard to imagine a mathematician setting out to write an e-less textbook, or being able to do so.

the condensed idea
The most natural
of numbers

07 Infinity

How big is infinity? The short answer is that ∞ (the symbol for infinity) is very big. Think of a straight line with larger and larger numbers lying along it and the line stretching 'off to infinity'. For every huge number produced, say 10^{1000}, there is always a bigger one, such as $10^{1000} + 1$.

This is a traditional idea of infinity, with numbers marching on forever. Mathematics uses infinity in any which way, but care has to be taken in treating infinity like an ordinary number. It is not.

Counting The German mathematician Georg Cantor gave us an entirely different concept of infinity. In the process, he single-handedly created a theory which has driven much of modern mathematics. The idea on which Cantor's theory depends has to do with a primitive notion of counting, simpler than the one we use in everyday affairs.

Imagine a farmer who didn't know about counting with numbers. How would he know how many sheep he had? Simple – when he lets his sheep out in the morning he can tell whether they are all back in the evening by pairing each sheep with a stone from a pile at the gate of his field. If there is a sheep missing there will be a stone left over. Even without using numbers, the farmer is being very mathematical. He is using the idea of a one-to-one correspondence between sheep and stones. This primitive idea has some surprising consequences.

Cantor's theory involves *sets* (a set is simply a collection of objects). For example **N** = {1, 2, 3, 4, 5, 6, 7, 8, . . .} means the set of (positive) whole numbers. Once we have a set, we can talk about subsets, which are smaller sets within the larger set. The most obvious subsets connected with our example **N** are the subsets **O** = {1, 3, 5, 7, . . .} and **E** = {2, 4, 6, 8, . . .}, which are the sets of the odd and even numbers respectively. If we were to ask 'is there the same number of odd numbers as even numbers?' what would be our answer? Though

timeline

we cannot do this by counting the elements in each set and comparing answers, the answer would still surely be 'yes'. What is this confidence based on? – probably something like 'half the whole numbers are odd and half are even'. Cantor would agree with the answer, but would give a different reason. He would say that every time we have an odd number, we have an even 'mate' next to it. The idea that both sets **O** and **E** have the same number of elements is based on the pairing of each odd number with an even number:

O: 1 3 5 7 9 11 13 15 17 19 21...

E: 2 4 6 8 10 12 14 16 18 20 22...

If we were to ask the further question 'is there the same number of whole numbers as *even* numbers?' the answer might be 'no', the argument being that the set **N** has twice as many numbers as the set of even numbers on its own.

The notion of 'more' though, is rather hazy when we are dealing with sets with an indefinite number of elements. We could do better with the one-to-one correspondence idea. Surprisingly, there is a one-to-one correspondence between **N** and the set of even numbers **E**:

N: 1 2 3 4 5 6 7 8 9 10 11...

E: 2 4 6 8 10 12 14 16 18 20 22...

We make the startling conclusion that there is the 'same number' of whole numbers as even numbers! This flies right in the face of the 'common notion' declared by the ancient Greeks; the beginning of Euclid of Alexandria's *Elements* text says that 'the whole is greater than the part'.

Cardinality The number of elements in a set is called its 'cardinality'. In the case of the sheep, the cardinality recorded by the farmer's accountants is 42. The cardinality of the set $\{a, b, c, d, e\}$ is 5 and this is written as $card\{a, b, c, d, e\} = 5$. So cardinality is a measure of the 'size' of a set. For the cardinality of

1874

1960s

Cantor treats the notion of infinity rigorously, specifying different orders of infinity

Abraham Robinson devises a non-standard arithmetic based on the notion of the infinitesimal

the whole numbers **N**, and any set in a one-to-one correspondence with **N**, Cantor used the symbol $\aleph_0$ ($\aleph$ or 'aleph' is from the Hebrew alphabet; the symbol $\aleph_0$ is read as 'aleph nought'). So, in mathematical language, we can write $card(\mathbf{N}) = card(\mathbf{O}) = card(\mathbf{E}) = \aleph_0$.

Any set which can be put into a one-to-one correspondence with **N** is called a 'countably infinite' set. Being countably infinite means we can write the elements of the set down in a list. For example, the list of odd numbers is simply 1, 3, 5, 7, 9, . . . and we know which element is first, which is second, and so on.

Are the fractions countably infinite?
The set of fractions **Q** is a larger set than **N** in the sense that **N** can be thought of as a subset of **Q**. Can we write all the elements of **Q** down in a list? Can we devise a list so that every fraction (including negative ones) is somewhere in it? The idea that such a big set could be put in a one-to-one correspondence with **N** seems impossible. Nevertheless it can be done.

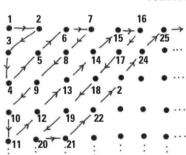

The way to begin is to think in two-dimensional terms. To start, we write down a row of all the whole numbers, positive and negative alternately. Beneath that we write all the fractions with 2 as denominator but we omit those which appear in the row above (like $\frac{4}{2} = 3$). Below this row we write those fractions which have 3 as denominator, again omitting those which have already been recorded. We continue in this fashion, of course never ending, but knowing exactly where every fraction appears in the diagram. For example, $^{209}\!/_{67}$ is in the 67th row, around 200 places to the right of $\frac{1}{67}$.

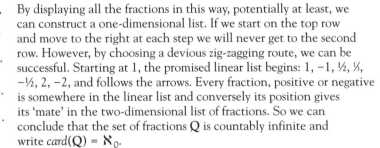

By displaying all the fractions in this way, potentially at least, we can construct a one-dimensional list. If we start on the top row and move to the right at each step we will never get to the second row. However, by choosing a devious zig-zagging route, we can be successful. Starting at 1, the promised linear list begins: 1, −1, ½, ⅓, −½, 2, −2, and follows the arrows. Every fraction, positive or negative is somewhere in the linear list and conversely its position gives its 'mate' in the two-dimensional list of fractions. So we can conclude that the set of fractions **Q** is countably infinite and write $card(\mathbf{Q}) = \aleph_0$.

Listing the real numbers
While the set of fractions accounts for many elements on the real number line there are also real numbers like √2, *e* and π which are *not* fractions. These are the irrational numbers – they 'fill in the gaps' to give us the real number line **R**.

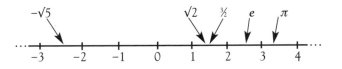

With the gaps filled in, the set **R** is referred to as the 'continuum'. So, how could we make a list of the real numbers? In a move of sheer brilliance, Cantor showed that even an attempt to put the real numbers *between 0 and 1* into a list is doomed to failure. This will undoubtedly come as a shock to people who are addicted to list-making, and they may indeed wonder how a set of numbers cannot be written down one after another.

Suppose you did not believe Cantor. You know that each number between 0 and 1 can be expressed as an extending decimal, for example, ½ = 0.500000000000000000... and ¼ = 0.31830988618379067153... and you would have to say to Cantor, 'here is my list of *all* the numbers between 0 and 1', which we'll call $r_1, r_2, r_3, r_4, r_5, \ldots$ If you could not produce one then Cantor would be correct.

Imagine Cantor looks at your list and he marks in **bold** the numbers on the diagonal:

$$r_1: \ 0.\boldsymbol{a_1}a_2a_3a_4a_5\ldots$$
$$r_2: \ 0.b_1\boldsymbol{b_2}b_3b_4b_5\ldots$$
$$r_3: \ 0.c_1c_2\boldsymbol{c_3}c_4c_5\ldots$$
$$r_4: \ 0.d_1d_2d_3\boldsymbol{d_4}d_5\ldots$$

Cantor would have said, 'OK, but where is the number $x = x_1x_2x_3x_4x_5\ldots$ where x_1 differs from a_1, x_2 differs from b_2, x_3 differs from c_3 working our way down the diagonal?' His x differs from *every* number in your list in one decimal place and so it cannot be there. Cantor is right.

In fact, no list is possible for the set of real numbers **R**, and so it is a 'larger' infinite set, one with a 'higher order of infinity', than the infinity of the set of fractions **Q**. Big just got bigger.

the condensed idea
A shower of infinities

08 Imaginary numbers

We can certainly imagine numbers. Sometimes I imagine my bank account is a million pounds in credit and there's no question that would be an 'imaginary number'. But the mathematical use of imaginary is nothing to do with this daydreaming.

The label 'imaginary' is thought to be due to the philosopher and mathematician René Descartes, in recognition of curious solutions of equations which were definitely not ordinary numbers. Do imaginary numbers exist or not? This was a question chewed over by philosophers as they focused on the word imaginary. For mathematicians the existence of imaginary numbers is not an issue. They are as much a part of everyday life as the number 5 or π. Imaginary numbers may not help with your shopping trips, but go and ask any aircraft designer or electrical engineer and you will find they are vitally important. And by adding a real number and an imaginary number together we obtain what's called a 'complex number', which immediately sounds less philosophically troublesome. The theory of complex numbers turns on the square root of *minus 1*. So what number, when squared, gives −1?

If you take any non-zero number and multiply it by itself (square it) you always get a positive number. This is believable when squaring positive numbers but is it true if we square negative numbers? We can use −1 × −1 as a test case. Even if we have forgotten the school rule that 'two negatives make a positive' we may remember that the answer is either −1 or +1. If we thought −1 × −1 equalled −1 we could divide each side by −1 and end up with the conclusion that −1 = 1, which is nonsense. So we must conclude −1 × −1 = 1, which is positive. The same argument can be made for other negative numbers besides −1, and so, when any real number is squared the result can *never* be negative.

timeline

This caused a sticking point in the early years of complex numbers in the 16th century. When this was overcome, the answer liberated mathematics from the shackles of ordinary numbers and opened up vast fields of inquiry undreamed of previously. The development of complex numbers is the 'completion of the real numbers' to a naturally more perfect system.

The square root of –1 We have already seen that, restricted to the real number line,

there is no square root of −1 as the square of any number cannot be negative. If we continue to think of numbers only on the real number line, we might as well give up, continue to call them imaginary numbers, go for a cup of tea with the philosophers, and have nothing more to do with them. Or we could take the bold step of accepting $\sqrt{-1}$ as a new entity, which we denote by i.

> **Engineering $\sqrt{-1}$**
>
> Even engineers, a very practical breed, have found uses for complex numbers. When Michael Faraday discovered alternating current in the 1830s, imaginary numbers gained a physical reality. In this case the letter j is used to represent $\sqrt{-1}$ instead of i because i stands for electrical current.

By this single mental act, imaginary numbers do exist. What they are we do not know, but we believe in their existence. At least we know $i^2 = -1$. So in our new system of numbers we have all our old friends like the real numbers 1, 2, 3, 4, π, e, $\sqrt{2}$ and $\sqrt{3}$, with some new ones involving i such as $1 + 2i$, $-3 + i$, $2 + 3i$, $1 + i\sqrt{2}$, $\sqrt{3} + 2i$, $e + \pi i$ and so on.

This momentous step in mathematics was taken around the beginning of the 19th century, when we escaped from the one-dimensional number line into a strange new two-dimensional number plane.

Adding and multiplying Now that we have complex numbers in our mind, numbers with the form $a + bi$, what can we do with them? Just like real numbers, they can be added and multiplied together. We add them by adding their respective parts. So $2 + 3i$ added to $8 + 4i$ gives $(2 + 8) + (3 + 4)i$ with the result $10 + 7i$.

Multiplication is almost as straightforward. If we want to multiply $2 + 3i$ by $8 + 4i$ we first multiply each pair of symbols together

$$(2 + 3i) \times (8 + 4i) = (2 \times 8) + (2 \times 4i) + (3i \times 8) + (3i \times 4i)$$

and add the resulting terms, 16, 8i, 24i and 12i² (in this last term, we replace i^2 by −1), together. The result of the multiplication is therefore (16 − 12) + (8i + 24i) which is the complex number 4 + 32i.

With complex numbers, all the ordinary rules of arithmetic are satisfied. Subtraction and division are always possible (except by the complex number 0 + 0i, but this was not allowed for zero in real numbers either). In fact the complex numbers enjoy all the properties of the real numbers save one. We cannot split them into positive ones and negative ones as we could with the real numbers.

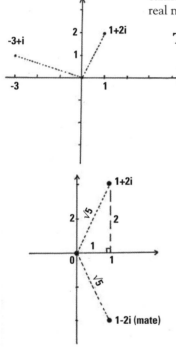

The Argand diagram The two-dimensionality of complex numbers is clearly seen by representing them on a diagram. The complex numbers −3 + i and 1 + 2i can be drawn on what we call an Argand diagram: This way of picturing complex numbers was named after Jean Robert Argand, a Swiss mathematician, though others had a similar notion at around the same time.

Every complex number has a 'mate' officially called its 'conjugate'. The mate of 1 + 2i is 1 − 2i found by reversing the sign in front of the second component. The mate of 1 − 2i, by the same token, is 1 + 2i, so that is true mateship.

Adding and multiplying mates together always produces a real number. In the case of adding 1 + 2i and 1 −2i we get 2, and multiplying them we get 5. This multiplication is more interesting. The answer 5 is the square of the 'length' of the complex number 1 + 2i and this equals the length of its mate. Put the other way, we could define the length of a complex number as:

$$length \ of \ w = \sqrt{(w \times mate \ of \ w)}$$

Checking this for −3 + i, we find that *length of* (−3 + i) = √(−3 + i × −3 − i) = √(9 + 1) and so the length of (−3 + i) = √10.

The separation of the complex numbers from mysticism owes much to Sir William Rowan Hamilton, Ireland's premier mathematician in the 19th century. He recognized that i wasn't actually needed for the theory. It only acted as a placeholder and could be thrown away. Hamilton considered a complex number as an 'ordered pair' of real numbers (a, b), bringing out their two-dimensional quality and making no appeal to the mystical $\sqrt{-1}$. Shorn of i, addition becomes

$$(2, 3) + (8, 4) = (10, 7)$$

and, a little less obviously, multiplication is

$$(2, 3) \times (8, 4) = (4, 32)$$

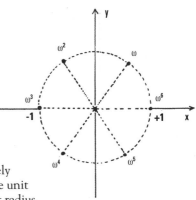

The completeness of the complex number system becomes clearer when we think of what are called 'the nth roots of unity' (for mathematicians 'unity' means 'one'). These are the solutions of the equation $z^n = 1$. Let's take $z^6 = 1$ as an example. There are the two roots $z = 1$ and $z = -1$ on the real number line (because $1^6 = 1$ and $(-1)^6 = 1$), but where are the others when surely there should be six? Like the two real roots, all of the six roots have unit length and are found on the circle centred at the origin and of unit radius.

More is true. If we look at $w = \frac{1}{2} + \frac{\sqrt{3}}{2}i$ which is the root in the first quadrant, the successive roots (moving in an anticlockwise direction) are $w^2, w^3, w^4, w^5, w^6 = 1$ and lie at the vertices of a regular hexagon. In general the n roots of unity will each lie on the circle and be at the corners or 'vertices' of a regular n-sided shape or polygon.

Extending complex numbers Once mathematicians had complex numbers they instinctively sought generalizations. Complex numbers are 2-dimensional, but what is special about 2? For years, Hamilton sought to construct 3-dimensional numbers and work out a way to add and multiply them but he was only successful when he switched to four dimensions. Soon afterwards these 4-dimensional numbers were themselves generalized to 8 dimensions (called Cayley numbers). Many wondered about 16-dimensional numbers as a possible continuation of the story – but 50 years after Hamilton's momentous feat, they were proved impossible.

the condensed idea
Unreal numbers with real uses

09 Primes

Mathematics is such a massive subject, criss-crossing all avenues of human enterprise, that at times it can appear overwhelming. Occasionally we have to go back to basics. This invariably means a return to the counting numbers, 1, 2, 3, 4, 5, 6, 7, 8, 9, 10, 11, 12, . . . Can we get more basic than this?

Well, $4 = 2 \times 2$ and so we can break it down into primary components. Can we break up any other numbers? Indeed, here are some more: $6 = 2 \times 3$, $8 = 2 \times 2 \times 2$, $9 = 3 \times 3$, $10 = 2 \times 5$, $12 = 2 \times 2 \times 3$. These are composite numbers for they are built up from the very basic ones 2, 3, 5, 7, . . . The 'unbreakable numbers' are the numbers 2, 3, 5, 7, 11, 13, . . . These are the prime numbers, or simply primes. A prime is a number which is only divisible by 1 and itself. You might wonder then if 1 itself is a prime number. According to this definition it should be, and indeed many prominent mathematicians in the past have treated 1 as a prime, but modern mathematicians start their primes with 2. This enables theorems to be elegantly stated. For us, too, the number 2 is the first prime.

For the first few counting numbers, we can underline the primes: 1, <u>2</u>, <u>3</u>, 4, <u>5</u>, 6, <u>7</u>, 8, 9, 10, <u>11</u>, 12, <u>13</u>, 14, 15, 16, <u>17</u>, 18, <u>19</u>, 20, 21, 22, <u>23</u>, . . . Studying prime numbers takes us back to the very basics of the basics. Prime numbers are important because they are the 'atoms' of mathematics. Like the basic chemical elements from which all other chemical compounds are derived, prime numbers can be built up to make mathematical compounds.

The mathematical result which consolidates all this has the grand name of the 'prime-number decomposition theorem'. This says that every whole number greater than 1 can be written by multiplying prime numbers in exactly one way. We saw that $12 = 2 \times 2 \times 3$ and there is no other way of doing it with prime components. This is often written in the power notation: $12 = 2^2 \times 3$. As another example, 6,545,448 can be written, $2^3 \times 3^5 \times 7 \times 13 \times 37$.

timeline

300BC	230BC
Euclid's *Elements* gives a proof that there are infinitely many prime numbers	Eratosthenes of Cyrene describes a method for sieving out prime numbers from the whole numbers

Discovering primes Unhappily there are no set formulae for identifying primes, and there seems to be no pattern in their appearances among the whole numbers. One of the first methods for finding them was developed by a younger contemporary of Archimedes who spent much of his life in Athens, Erastosthenes of Cyrene. His precise calculation of the length of the equator was much admired in his own time. Today he's noted for his sieve for finding prime numbers. Erastosthenes imagined the counting numbers stretched out before him. He underlined **2** and struck out all multiples of 2. He then moved to 3, underlined it and struck out all multiples of 3. Continuing in this way, he sieved out all the composites. The underlined numbers left behind in the sieve were the primes.

0	1	2	3	4	5	6	7	8	9
10	11	12	13	14	15	16	17	18	19
20	21	22	23	24	25	26	27	28	29
30	31	32	33	34	35	36	37	38	39
40	41	42	43	44	45	46	47	48	49
50	51	52	53	54	55	56	57	58	59
60	61	62	63	64	65	66	67	68	69
70	71	72	73	74	75	76	77	78	79
80	81	82	83	84	85	86	87	88	89
90	91	92	93	94	95	96	97	98	99

So we can predict primes, but how do we decide whether a given number is a prime or not? How about 19,071 or 19,073? Except for the primes 2 and 5, a prime number must end in a 1, 3, 7 or 9 but this requirement is not enough to *make* that number a prime. It is difficult to know whether a large number ending in 1, 3, 7 or 9 is a prime or not without trying possible factors. By the way, $19,071 = 3^2 \times 13 \times 163$ is not a prime, but 19,073 is.

Another challenge has been to discover any patterns in the distribution of the primes. Let's see how many primes there are in each segment of 100 between 1 and 1000.

Range	1–100	101–200	201–300	301–400	401–500	501–600	601–700	701–800	801–900	901–1000	1–1000
Number of primes	25	21	16	16	17	14	16	14	15	14	168

In 1792, when only 15 years old, Carl Friedrich Gauss suggested a formula $P(n)$ for estimating the number of prime numbers less than a given number n (this is now called the prime number theorem). For $n = 1000$ the formula gives the approximate value of 172. The actual number of primes, 168, is less than this

AD**1742**

Goldbach speculates that every even number (more than 2) is the sum of two primes

1896

The prime number theorem on the distribution of primes is proved

1966

Chen Jingrun almost confirms the Goldbach conjecture

estimate. It had always been assumed this was the case for any value of n, but the primes often have surprises in store and it has been shown that for $n = 10^{371}$ (a huge number written long hand as a 1 with 371 trailing 0s) the actual number of primes *exceeds* the estimate. In fact, in some regions of the counting numbers the difference between the estimate and the actual number oscillates between less and excess.

How many? There are infinitely many prime numbers. Euclid stated in his *Elements* (Book 9, Proposition 20) that 'prime numbers are more than any assigned multitude of prime numbers'. Euclid's beautiful proof goes like this:

> Suppose that P is the largest prime, and consider the number $N = (2 \times 3 \times 5 \times \ldots \times P) + 1$. Either N is prime or it is not. If N is prime we have produced a prime greater than P which is a contradiction to our supposition. If N is not a prime it must be divisible by some prime, say p, which is one of 2, 3, 5, . . ., P. This means that p divides $N - (2 \times 3 \times 5 \times \ldots \times P)$. But this number is equal to 1 and so p divides 1. This cannot be since all primes are greater than 1. Thus, whatever the nature of N, we arrive at a contradiction. Our original assumption of there being a largest prime P is therefore false. *Conclusion:* the number of primes is limitless.

Though primes 'stretch to infinity' this fact has not prevented people striving to find the largest known prime. One which has held the record recently is the enormous Mersenne prime $2^{24036583} - 1$, which is approximately $10^{7235732}$ or a number starting with 1 followed by 7,235,732 trailing zeroes.

The unknown Outstanding unknown areas concerning primes are the 'Twin primes problem' and the famous 'Goldbach conjecture'.

Twin primes are pairs of consecutive primes separated only by an even number. The twin primes in the range from 1 to 100 are 3, 5; 5, 7; 11, 13; 17, 19; 29, 31; 41, 43; 59, 61; 71, 73. On the numerical front, it is known there are 27,412,679 twins less than 10^{10}. This means the even numbers with twins, like 12 (having twins 11, 13), constitute only 0.274% of the numbers in this range. Are there an infinite number of twin primes? It would be curious if there were not, but no one has so far been able write down a proof of this.

Christian Goldbach conjectured that:

> *Every even number greater than 2 is the sum of two prime numbers.*

For instance, 42 is an even number and we can write it as 5 + 37. The fact that we can also write it as 11 + 31, 13 + 29 or 19 + 23 is beside the point – all we need is *one* way. The conjecture is true for a huge range of numbers – but it has never been proved in general. However, progress has been made, and some have a feeling that a proof is not far off. The Chinese mathematician Chen Jingrun made a great step. His theorem states that every sufficiently large even number can be written as the sum of two primes *or* the sum of a prime and a *semi*-prime (a number which is the multiplication of two primes).

The great number theorist Pierre de Fermat proved that primes of the form $4k + 1$ are expressible as the sum of two squares in exactly one way (e.g. $17 = 1^2 + 4^2$), while those of the form $4k + 3$ (like 19) cannot be written as the sum of two squares at all. Joseph Lagrange also proved a famous mathematical theorem about square powers: *every* positive whole number is the sum of four squares. So, for example, $19 = 1^2 + 1^2 + 1^2 + 4^2$. Higher powers have been explored and books filled with theorems, but many problems remain.

The number of the numerologist

One of the most challenging areas of number theory concerns 'Waring's problem'. In 1770 Edward Waring, a professor at Cambridge, posed problems involving writing whole numbers as the addition of powers. In this setting the magic arts of numerology meet the clinical science of mathematics in the shape of primes, sums of squares and sums of cubes. In numerology, take the unrivalled cult number 666, the 'number of the beast' in the biblical book of *Revelation*, and which has some unexpected properties. It is the sum of the squares of the first 7 primes:

$$666 = 2^2 + 3^2 + 5^2 + 7^2 + 11^2 + 13^2 + 17^2$$

Numerologists will also be keen to point out that it is the sum of palindromic cubes and, if that is not enough, the keystone 6^3 in the centre is shorthand for $6 \times 6 \times 6$:

$$666 = 1^3 + 2^3 + 3^3 + 4^3 + 5^3 + 6^3 + 5^3 + 4^3 + 3^3 + 2^3 + 1^3$$

The number 666 is truly the 'number of the numerologist'.

We described the prime numbers as the 'atoms of mathematics'. But 'surely,' you might say, 'physicists have gone beyond atoms to even more fundamental units, like quarks. Has mathematics stood still?' If we limit ourselves to the counting numbers, 5 is a prime number and will always be so. But Gauss made a far-reaching discovery, that for some primes, like 5, $5 = (1 - 2i) \times (1 + 2i)$ where $i = \sqrt{-1}$ of the imaginary number system. As the product of two Gaussian integers, 5 and numbers like it are not as unbreakable as was once supposed.

the condensed idea
The atoms of mathematics

10 Perfect numbers

In mathematics the pursuit of perfection has led its aspirants to different places. There are perfect squares, but here the term is not used in an aesthetic sense. It's more to warn you that there are imperfect squares in existence. In another direction, some numbers have few divisors and some have many. But, like the story of the three bears, some numbers are 'just right'. When the addition of the divisors of a number equals the number itself it is said to be perfect.

The Greek philosopher Speusippus, who took over the running of the Academy from his uncle Plato, declared that the Pythagoreans believed that 10 had the right credentials for perfection. Why? Because the number of prime numbers between 1 and 10 (namely 2, 3, 5, 7) equalled the non-primes (4, 6, 8, 9) and this was the smallest number with this property. Some people have a strange idea of perfection.

It seems the Pythagoreans actually had a richer concept of a perfect number. The mathematical properties of perfect numbers were delineated by Euclid in the *Elements* and studied in depth by Nicomachus 400 years later, leading to amicable numbers and even sociable numbers. These categories were defined in terms of the relationships between them and their divisors. At some point they came up with the theory of superabundant and deficient numbers and this led them to their concept of perfection.

Whether a number is superabundant is determined by its divisors and makes a play on the connection between multiplication and addition. Take the number 30 and consider its divisors, that is all the numbers which divide into it exactly and which are *less* than 30. For such a small number as 30 we can see the divisors are 1, 2, 3, 5, 6, 10 and 15. Totalling up these divisors we get 42. The

timeline

525BC	**300**BC	AD**100**
The Pythagoreans are associated with both perfect and abundant numbers	Book 9 of Euclid's *Elements* discusses perfect numbers	Nicomachus of Gerasa gives a classification of numbers based on perfect numbers

Rank	1	2	3	4	5	6	7
Perfect number	6	28	496	8128	33,550,336	8,589,869,056	137,438,691,328

The first few perfect numbers

number 30 is superabundant because the addition of its divisors (42) is bigger than the number 30 itself.

A number is deficient if the opposite is true – if the sum of its divisors is less than itself. So the number 26 is deficient because its divisors 1, 2 and 13 add up to only 16, which is less than 26. Prime numbers are very deficient because the sum of their divisors is always just 1.

A number that is neither superabundant nor deficient is perfect. The addition of the divisors of a perfect number equal the number itself. The first perfect number is 6. Its divisors are 1, 2, 3 and when we add them up, we get 6. The Pythagoreans were so enchanted with the number 6 and the way its parts fitted together that they called it 'marriage, health and beauty'. There is another story connected with 6 told by St Augustine (354–430). He believed that the perfection of 6 existed before the world came into existence and that the world was created in 6 days *because* the number was perfect.

The next perfect number is 28. Its divisors are 1, 2, 4, 7 and 14 and, when we add them up, we get 28. These first two perfect numbers, 6 and 28, are rather special in perfect number lore for it can be proved that every even perfect number ends in a 6 or a 28. After 28, you have wait until 496 for the next perfect number. It is easy to check it really is the sum of its divisors: 496 = 1 + 2 + 4 + 8 + 16 + 31 + 62 + 124 +248. For the next perfect numbers we have to start going into the numerical stratosphere. The first five were known in the 16th century, but we still don't know if there is a largest one, or whether they go marching on without limit. The balance of opinion suggests that they, like the primes, go on for ever.

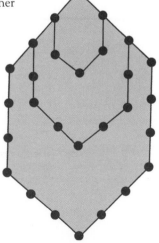

The Pythagoreans were keen on geometrical connections. If we have a perfect number of beads, they can be arranged around a hexagonal necklace. In the case of 6 this is the simple hexagon with beads placed at its corners, but for higher perfect numbers we have to add in smaller sub-necklaces within the large one.

1603

Pietro Cataldi finds the sixth and seventh perfect numbers, $2^{16}(2^{17} - 1)$ = 8,589,869,056 and $2^{18}(2^{19} - 1)$ = 137,438,691,328

2006

The great prime search project finds the 44th Mersenne prime (with almost ten million digits) and yet another new perfect number can be generated

The power	Result	Take away 1 (Mersenne number)	Prime number?
2	4	3	prime
3	8	7	prime
4	16	15	not prime
5	32	31	prime
6	64	63	not prime
7	128	127	prime
8	256	255	not prime
9	512	511	not prime
10	1,024	1,023	not prime
11	2,048	2,047	not prime
12	4,096	4,095	not prime
13	8,192	8,191	prime
14	16,384	16,383	not prime
15	32,768	32,767	not prime

Mersenne numbers The key to constructing perfect numbers is a collection of numbers named after Father Marin Mersenne, a French monk who studied at a Jesuit college with René Descartes. Both men were interested in finding perfect numbers. Mersenne numbers are constructed from powers of 2, the doubling numbers 2, 4, 8, 16, 32, 64, 128, 256, . . ., and then subtracting a single 1. A Mersenne number is a number of the form $2^n - 1$. While they are always odd, they are not always prime. But it is those Mersenne numbers that are also prime that can be used to construct perfect numbers.

Mersenne knew that if the power was *not* a prime number, then the Mersenne number could not be a prime number either, accounting for the non-prime powers 4, 6, 8, 9, 10, 12, 14 and 15 in the table. The Mersenne numbers could only be prime if the power was a prime number, but was that enough? For the first few cases, we do get 3, 7, 31 and 127, all of which are prime. So is it generally true that a Mersenne number formed with a prime power should be prime as well?

Many mathematicians of the ancient world up to about the year 1500 thought this was the case. But primes are not constrained by simplicity, and it was found that for the power 11 (a prime number), $2^{11} - 1 = 2047 = 23 \times 89$ and consequently it is not a prime number. There seems to be no rule. The

Just good friends

The hard-headed mathematician is not usually given to the mystique of numbers but numerology is not yet dead. The amicable numbers came after the perfect numbers though they may have been known to the Pythagoreans. Later they became useful in compiling romantic horoscopes where their mathematical properties translated themselves into the nature of the ethereal bond. The two numbers 220 and 284 are amicable numbers. Why so? Well, the divisors of 220 are 1, 2, 4, 5, 10, 11, 20, 22, 44, 55 and 110 and if you add them up you get 284. You've guessed it. If you figure out the divisors of 284 and add them up, you get 220. That's true friendship.

Mersenne numbers $2^{17} - 1$ and $2^{19} - 1$ are both primes, but $2^{23} - 1$ is not a prime, because

$$2^{23} - 1 = 8,388,607 = 47 \times 178,481$$

Construction work A combination of Euclid and Euler's work provides a formula which enables even perfect numbers to be generated: n is an even perfect number if and only if $n = 2^{p-1}(2^p - 1)$ where $2^p - 1$ is a Mersenne prime.

For example, $6 = 2^1(2^2 - 1)$, $28 = 2^2(2^3 - 1)$ and $496 = 2^4(2^5 - 1)$. This formula for calculating even perfect numbers means we can generate them if we can find Mersenne primes. The perfect numbers have challenged both people and machines and will continue to do so in a way which earlier practitioners had not envisaged. Writing at the beginning of the 19th century, the table maker Peter Barlow thought that no one would go beyond the calculation of Euler's perfect number

$$2^{30}(2^{31} - 1) = 2,305,843,008,139,952,128$$

as there was little point. He could not foresee the power of modern computers or mathematicians' insatiable need to meet new challenges.

Odd perfect numbers No one knows if an odd perfect number will ever be found. Descartes did not think so but experts can be wrong. The English mathematician James Joseph Sylvester declared the existence of an odd perfect number 'would be little short of a miracle' because it would have to satisfy so many conditions. It's little surprise Sylvester was dubious. It is one of the oldest problems in mathematics, but if an odd perfect number does exist quite a lot is already known about it. It would need to have at least 8 distinct prime divisors, one of which is greater than a million, while it would have to be at least 300 digits long.

Mersenne Primes

Finding Mersenne primes is not easy. Many mathematicians over the centuries have added to the list, which has a chequered history built on a combination of error and correctness. The great Leonhard Euler contributed the eighth Mersenne prime, $2^{31} - 1 = 2,147,483,647$, in 1732. Finding the 23rd Mersenne prime, $2^{11213} - 1$, in 1963 was a source of pride for the mathematics department at the University of Illinois, who announced it to the world on their university postage stamp. But with powerful computers the Mersenne prime industry had moved on and in the late 1970s high school students Laura Nickel and Landon Noll jointly discovered the 25th Mersenne prime, and Noll the 26th Mersenne prime. To date 45 Mersenne primes have been discovered.

the condensed idea
The mystique of numbers

11 Fibonacci numbers

In *The Da Vinci Code*, the author Dan Brown made his murdered curator Jacques Saunière leave behind the first eight terms of a sequence of numbers as a clue to his fate. It required the skills of cryptographer Sophie Neveu to reassemble the numbers 13, 3, 2, 21, 1, 1, 8 and 5 to see their significance. Welcome to the most famous sequence of numbers in all of mathematics.

The Fibonacci sequence of whole numbers is:

1, 1, 2, 3, 5, 8, 13, 21, 34, 55, 89, 144, 233, 377, 610, 987, 1597, 2584, . . .

The sequence is widely known for its many intriguing properties. The most basic – indeed the characteristic feature which defines them – is that every term is the addition of the previous two. For example 8 = 5 + 3, 13 = 8 + 5, . . ., 2584 = 1587 + 987, and so on. All you have to remember is to begin with the two numbers 1 and 1 and you can generate the rest of the sequence on the spot. The Fibonacci sequence is found in nature as the number of spirals formed from the number of seeds in the spirals in sunflowers (for example, 34 in one direction, 55 in the other), and the room proportions and building proportions designed by architects. Classical musical composers have used it as an inspiration, with Bartók's Dance Suite believed to be connected to the sequence. In contemporary music Brian Transeau (aka BT) has a track in his album *This Binary Universe* called 1.618 as a salute to the ultimate ratio of the Fibonacci numbers, a number we shall discuss a little later.

Origins The Fibonacci sequence occurred in the *Liber Abaci* published by Leonardo of Pisa (Fibonacci) in 1202, but these numbers were probably known in India before that. Fibonacci posed the following problem of rabbit generation:

timeline

AD 1202

Leonardo of Pisa publishes the
Liber Abaci and Fibonacci
numbers

1724

Daniel Bernoulli expresses
numbers of the Fibonacci s
in terms of the golden ratio

Mature rabbit pairs generate young rabbit pairs each month. At the beginning of the year there is one young rabbit pair. By the end of the first month they will have matured, by the end of the second month the mature pair is still there and they will have generated a young rabbit pair. The process of maturing and generation continues. Miraculously none of the rabbit pairs die.

O = young pair

● = mature pair

Fibonacci wanted to know how many rabbit pairs there would be at the end of the year. The generations can be shown in a 'family tree'. Let's look at the number of pairs at the end of May (the fifth month). We see the number of pairs is 8. In this layer of the family tree the left-hand group

● O ● O

is a duplicate of the whole row above, and the right-hand group

● O ●

is a duplicate of the row above that. This shows that the birth of rabbit pairs follows the basic Fibonacci equation:

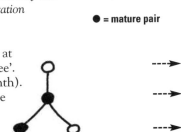

The rabbit population

number after n months = number after $(n-1)$ month
+ number after $(n-2)$ months

Properties   Let's see what happens if we add the terms of the sequence:

$$1 + 1 = 2$$
$$1 + 1 + 2 = 4$$
$$1 + 1 + 2 + 3 = 7$$
$$1 + 1 + 2 + 3 + 5 = 12$$
$$1 + 1 + 2 + 3 + 5 + 8 = 20$$
$$1 + 1 + 2 + 3 + 5 + 8 + 13 = 33$$
$$\cdots$$

The result of each of these sums will form a sequence as well, which we can place under the original sequence, but shifted along:

923

artók composes his 'Dance
uite', believed to be inspired by
e Fibonacci numbers

1963

The *Fibonacci Quarterly*, a
journal devoted to the number
theory of the Fibonacci
sequence, is founded

2007

Sculptor Peter Randall-Page creates
the 70 tonne sculpture 'Seed' based
on the Fibonacci sequence for the
Eden Project in Cornwall, UK

Fibonacci	1	1	2	3	5	8	13	21	34	55	89 . . .

Addition		2	4	7	12	20	33	54	88 . . .

The addition of *n* terms of the Fibonacci sequence turns out to be 1 less than the next but one Fibonacci number. If you want to know the answer to the addition of 1 + 1 + 2 + . . . + 987, you just subtract 1 from 2584 to get 2583. If the numbers are added alternately by missing out terms, such as 1 + 2 + 5 + 13 + 34, we get the answer 55, itself a Fibonacci number. If the other alternation is taken, such as 1 + 3 + 8 + 21 + 55, the answer is 88 which is a Fibonacci number less 1.

The squares of the Fibonacci sequence numbers are also interesting. We get a new sequence by multiplying each Fibonacci number by itself and adding them.

Fibonacci	1	1	2	3	5	8	_13_	_21_	34	55 . . .

Squares	1	1	4	9	25	64	169	441	1156	3025 . . .

Addition of squares	1	2	6	15	40	104	_273_	714	1870	4895 . . .

In this case, adding up all the squares up to the *n*th member is the same as multiplying the *n*th member of the original Fibonacci sequence by the next one to this. For example,

$$1 + 1 + 4 + 9 + 25 + 64 + 169 = 273 = 13 \times 21$$

Fibonacci numbers also occur when you don't expect them. Let's imagine we have a purse containing a mix of £1 and £2 coins. What if we want to count the number of ways the coins can be taken from the purse to make up a particular amount expressed in pounds. In this problem the order of actions is important. The value of £4, as we draw the coins out of the purse, can be any of the following ways, 1 + 1 + 1 + 1; 2 + 1 +1; 1 + 2 + 1; 1 + 1 + 2; and 2 + 2. There are 5 ways in all – and this corresponds to the fifth Fibonacci number. If you take out £20 there are 6,765 ways of taking the £1 and £2 coins out, corresponding to the 21st Fibonacci number! This shows the power of simple mathematical ideas.

The golden ratio If we look at the ratio of terms formed from the Fibonacci sequence by dividing a term by its preceding term we find out another remarkable property of the Fibonacci numbers. Let's do it for a few terms 1, 1, 2, 3, 5, 8, 13, 21, 34, 55.

1/1	2/1	3/2	5/3	8/5	13/8	21/13	34/21	55/34
1.000	2.000	1.500	1.333	1.600	1.625	1.615	1.619	1.617

Pretty soon the ratios approach a value known as the golden ratio, a famous number in mathematics, designated by the Greek letter ϕ. It takes its place amongst the top mathematical constants like π and e, and has the exact value

$$\phi = \frac{1+\sqrt{5}}{2}$$

and this can be approximated to the decimal 1.618033988... With a little more work we can show that each Fibonacci number can be written in terms of ϕ.

Despite the wealth of knowledge known about the Fibonacci sequence, there are still many questions left to answer. The first few prime numbers in the Fibonacci sequence are 2, 3, 5, 13, 89, 233, 1597 – but we don't know if there are infinitely many primes in the Fibonacci sequence.

Family resemblances The Fibonacci sequence holds pride of place in a wide ranging family of similar sequences. A spectacular member of the family is one we may associate with a cattle population problem. Instead of Fibonacci's rabbit pairs which transform in one month from young pair to mature pair which then start breeding, there is an intermediate stage in the maturation process as cattle pairs progress from young pairs to immature pairs and then to mature pairs. It is only the mature pairs which can reproduce. The cattle sequence is:

1, 1, 1, 2, 3, 4, 6, 9, 13, 19, 28, 41, 60, 88, 129, 189, 277, 406, 595, . . .

Thus the generation skips a value so for example, 41 = 28 + 13 and 60 = 41 + 19. This sequence has similar properties to the Fibonacci sequence. For the cattle sequence the ratios obtained by dividing a term by its preceding term approach the limit denoted by the Greek letter psi, written ψ, where

$$\psi = 1.46557123187676802665. . .$$

This is known as the 'supergolden ratio'.

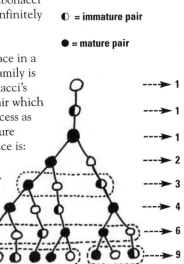

O = young pair

◖ = immature pair

● = mature pair

---➤ 1
---➤ 1
---➤ 1
---➤ 2
---➤ 3
---➤ 4
---➤ 6
---➤ 9

The cattle population

the condensed idea
The Da Vinci Code unscrambled

12 Golden rectangles

Rectangles are all around us – buildings, photographs, windows, doors, even this book. Rectangles are present within the artists' community – Piet Mondrian, Ben Nicholson and others, who progressed to abstraction, all used one sort or another. So which is the most beautiful of all? Is it a long thin 'Giacometti rectangle' or one that is almost a square? Or is it a rectangle in between these extremes?

Does the question even make sense? Some think so, and believe particular rectangles are more 'ideal' than others. Of these, perhaps the golden rectangle has found greatest favour. Amongst all the rectangles one could choose for their different proportions – for that is what it comes down to – the golden rectangle is a very special one which has inspired artists, architects and mathematicians. Let's look at some other rectangles first.

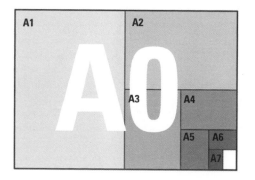

Mathematical paper If we take a piece of A4 paper, whose dimensions are a short side of 210 mm and a long side of 297 mm, the length-to-width ratio will be 297/210 which is approximately 1.4142. For any international A-size paper with short side equal to b, the longer side will always be $1.4142 \times b$. So for A4, $b = 210$ mm, while for A5, $b = 148$ mm. The A-formulae system used for paper sizes has a highly desirable property, one that does not occur for arbitrary paper sizes. If an A-size piece of paper is folded about the middle, the two smaller rectangles formed are directly in proportion to the larger rectangle. They are two smaller versions of the *same* rectangle.

In this way, a piece of A4 folded into two pieces generates two pieces of A5. Similarly a piece of A5-size paper generates two pieces of A6. In the other direction, a sheet of A3 paper is made up of two pieces of A4. The smaller the number on the A-size the larger the piece of paper. How did we know that the particular number 1.4142 would do the trick? Let's fold a rectangle, but this time let's make it one where we don't know the length of its longer side. If we take the breadth of a rectangle to be 1 and we write the length of the longer side as x, then the length-to-width ratio is $x/1$. If we now fold the rectangle, the length-to-width ratio of the smaller rectangle is $1/\frac{1}{2}x$, which is the same as $2/x$. The point of A sizes is that our two ratios must stand for the same proportion, so we get an equation $x/1 = 2/x$ or $x^2 = 2$. The true value of x is therefore $\sqrt{2}$ which is approximately by 1.4142.

Mathematical gold The golden rectangle is different, but only *slightly* different. This time the rectangle is folded along the line RS in the diagram so that the points MRSQ make up the corners of a *square*.

The key property of the golden rectangle is that the rectangle left over, RNPS, is proportional to the large rectangle – what is left over should be a mini-replica of the large rectangle.

As before, we'll say the breadth MQ = MR of the large rectangle is 1 unit of length while we'll write the length of the longer side MN as x. The length-to-width ratio is again $x/1$. This time the breadth of the smaller rectangle RNPS is MN – MR, which is $x - 1$ so the length-to-width ratio of this rectangle is $1/(x-1)$. By equating them, we get the equation

$$\frac{x}{1} = \frac{1}{x-1}$$

which can be multiplied out to give $x^2 = x + 1$. An approximate solution is 1.618. We can easily check this. If you type 1.618 into a calculator and multiply it by itself you get 2.618 which is the same as $x + 1 = 2.618$. This number is the famous golden ratio and is designated by the Greek letter phi, ϕ. Its definition and approximation is given by

$$\phi = \frac{1+\sqrt{5}}{2} = 1.61803398874989484820...$$

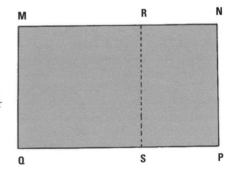

aciola publishes *The Divine Proportion*

Fechner writes on psychological experiments to determine the proportions of the most 'aesthetic' rectangle

The International Organization for Standardization (ISO) defines the A paper size

and this number is related to the Fibonacci sequence and the rabbit problem (see page 44).

Going for gold Now let's see if we can build a golden rectangle. We'll begin with our square MQSR with sides equal to 1 unit and mark the midpoint of QS as O. The length OS = ½, and so by Pythagoras's theorem (see page 84) in the triangle ORS, $OR = \sqrt{\left(\frac{1}{2}\right)^2 + 1^2} = \frac{\sqrt{5}}{2}$

Using a pair of compasses centred on O, we can draw the arc RP and we'll find that OP = OR = $\sqrt{5}$⁄2 . So we end up with

$$QP = \frac{1}{2} + \frac{\sqrt{5}}{2} = \phi$$

which is what we wanted: the 'golden section' or the side of the golden rectangle.

History Much is claimed of the golden ratio ϕ. Once its appealing mathematical properties are realized it is possible to see it in unexpected places, even in places where it is not. More than this is the danger of claiming the golden ratio was there before the artefact – that musicians, architects and artists had it in mind at the point of creation. This foible is termed 'golden numberism'. The progress from numbers to general statements without other evidence is a dangerous argument to make.

Take the Parthenon in Athens. At its time of construction the golden ratio was certainly known but this does not mean that the Parthenon was based on it. Sure, in the front view of the Parthenon the ratio of the width to the height (including the triangular pediment) is 1.74 which is close to 1.618, but is it close enough to claim the golden ratio as a motivation? Some argue that the pediment should be left out of the calculation, and if this is done, the width-to-height ratio is actually the whole number 3.

In his 1509 book De divina proportione, Luca Pacioli 'discovered' connections between characteristics of God and properties of the proportion determined by ϕ. He christened it the 'divine proportion'. Pacioli was a Franciscan monk who wrote influential books on mathematics. By some he is regarded as the 'father of accounting' because he popularized the double-entry method of accounting used by Venetian merchants. His other claim to fame is that he taught mathematics to Leonardo da Vinci. In the Renaissance, the golden section achieved near mystical status – the astronomer Johannes Kepler described it as a mathematical 'precious jewel'. Later, Gustav Fechner, a German experimental psychologist, made thousands of measurements of rectangular shapes (playing cards, books, windows) and found the most commonly occurring ratio of their sides was close to ϕ.

Le Corbusier was fascinated by the rectangle as a central element in architectural design and by the golden rectangle in particular. He placed great emphasis on harmony and order and found this in mathematics. He saw architecture through the eyes of a mathematician. One of his planks was the 'modulator' system, a theory of proportions. In effect this was a way of generating streams of golden rectangles, shapes he used in his designs. Le Corbusier was inspired by Leonardo da Vinci who, in turn, had taken careful notes on the Roman architect Vitruvius, who set store by the proportions found in the human figure.

Other shapes There is also a 'supergolden rectangle' whose construction has similarities with the way the golden rectangle is constructed.

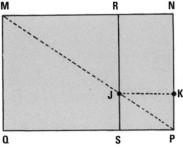

This is how we build the supergolden rectangle MQPN. As before, MQSR is a square whose side is of length 1. Join the diagonal MP and mark the intersection on RS as the point J. Then make a line JK that's parallel to RN with K on NP. We'll say the length RJ is y and the length MN is x. For any rectangle, RJ/MR = NP/MN (because triangles MRJ and MNP are similar), so $y/1 = 1/x$ which means $x \times y = 1$ and we say x and y are each other's 'reciprocal'. We get the supergolden rectangle by making the rectangle RJKN proportional to the original rectangle MQPN, that is $y/(x - 1) = x/1$. Using the fact that $xy = 1$, we can conclude that the length of the supergolden rectangle x is found by solving the 'cubic' equation $x^3 = x^2 + 1$, which is clearly similar to the equation $x^2 = x + 1$ (the equation that determines the golden rectangle). The cubic equation has one positive real solution ψ (replacing x with the more standard symbol ψ) whose value is

$$\psi = 1.46557123187676802665\ldots$$

the number associated with the cattle sequence (see page 47). Whereas the golden rectangle can be constructed by a straight edge and a pair of compasses, the supergolden rectangle cannot be made this way.

the condensed idea
Divine proportions

13 Pascal's triangle

The number 1 is important but what about 11? It is interesting too and so is 11 × 11 = 121, 11 × 11× 11 = 1331 and 11 × 11 × 11 × 11 = 14,641. Setting these out we get

11
121
1331
14,641

These are the first lines of Pascal's triangle. But where do we find it?

Throwing in $11^0 = 1$ for good measure, the first thing to do is forget the commas, and then introduce spaces between the numbers. So 14,641 becomes 1 4 6 4 1.

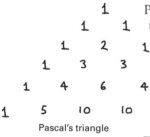

Pascal's triangle

Pascal's triangle is famous in mathematics for its symmetry and hidden relationships. In 1653 Blaise Pascal thought so and remarked that he could not possibly cover them all in one paper. The many connections of Pascal's triangle with other branches of mathematics have made it into a venerable mathematical object, but its origins can be traced back much further than this. In fact Pascal didn't invent the triangle named after him – it was known to Chinese scholars of the 13th century.

The Pascal pattern is generated from the top. Start with a 1 and place two 1s on either side of it in the next row down. To construct further rows we continue to place 1s on the ends of each row while the internal numbers are obtained by the sum of the two numbers immediately above. To obtain 6 in

timeline

c.500BC

Fragmentary evidence exists for Pascal's triangle in Sanscrit

c.AD1070

Omar Khayyam discovers the triangle, which in some countries is named after him

the fifth row for example, we add 3 + 3 from the row above. The English mathematician G.H. Hardy said 'a mathematician, like a painter or a poet, is a maker of patterns' and Pascal's triangle has patterns in spades.

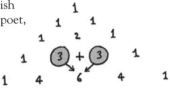

Links with algebra Pascal's triangle is founded on real mathematics. If we work out $(1 + x) \times (1 + x) \times (1 + x) = (1 + x)^3$, for example, we get $1 + 3x + 3x^2 + x^3$. Look closely and you'll see the numbers in front of the symbols in this expression match the numbers in the corresponding row of Pascal's triangle. The scheme followed is:

$$
\begin{array}{c}
(1 + x)^0 \qquad\qquad\qquad 1 \\
(1 + x)^1 \qquad\qquad\quad 1 \quad 1 \\
(1 + x)^2 \qquad\qquad 1 \quad 2 \quad 1 \\
(1 + x)^3 \qquad\quad 1 \quad 3 \quad 3 \quad 1 \\
(1 + x)^4 \qquad 1 \quad 4 \quad 6 \quad 4 \quad 1 \\
(1 + x)^5 \quad 1 \quad 5 \quad 10 \quad 10 \quad 5 \quad 1
\end{array}
$$

If we add up the numbers in any row of Pascal's triangle we always obtain a power of 2. For example in the fifth row down $1 + 4 + 6 + 4 + 1 = 16 = 2^4$. This can be obtained from the left-hand column above if we use $x = 1$.

Properties The first and most obvious property of Pascal's triangle is its symmetry. If we draw a vertical line down through the middle, the triangle has 'mirror symmetry' – it is the same to the left of the vertical line as to the right of it. This allows us to talk about plain 'diagonals', because a northeast diagonal will be the same as a northwest diagonal. Under the diagonal made up of 1s we have the diagonal made up of the counting numbers 1, 2, 3, 4, 5, 6, . . . Under that there are the triangular numbers, 1, 3, 6, 10, 15, 21, . . . (the numbers which can be made up of dots in the form of triangles). In the diagonal under that we have the tetrahedral numbers, 1, 4, 10, 20, 35, 56, . . . These numbers correspond to tetrahedra ('three-dimensional triangles', or, if you like, the number of cannon balls which can be placed on triangular bases of increasing sizes). And what about the 'almost diagonals'?

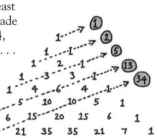

Almost diagonals in Pascal's triangle

If we add up the numbers in lines across the triangle (which are not rows or true diagonals), we get the sequence 1, 2, 5, 13, 34, . . . Each number is three times the previous one with the one before that

1303

Zhu Shijie defines Pascal's triangle and shows how to sum certain sequences

1664

Pascal's paper on the properties of the triangle is published posthumously

1714

Leibniz discusses the harmonic triangle

subtracted. For example $34 = 3 \times 13 - 5$. Based on this, the next number in the sequence will be $3 \times 34 - 13 = 89$. We have missed out the alternate 'almost diagonals', starting with $1, 1 + 2 = 3$, but these will give us the sequence $\underline{1}, \underline{3}, \underline{8}$, $\underline{21}, \underline{55}, \ldots$ and these are generated by the same '3 times minus 1' rule. We can therefore generate the next number in the sequence, as $3 \times 55 - 21 = \underline{144}$. But there's more. If we interleave these two sequences of 'almost diagonals' we get the Fibonacci numbers:

$$1, \underline{1}, 2, \underline{3}, 5, \underline{8}, 13, \underline{21}, 34, \underline{55}, 89, \underline{144}, \ldots$$

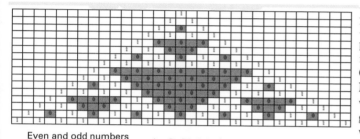

Even and odd numbers in Pascal's triangle

Pascal combinations The Pascal numbers answer some counting problems. Think about 7 people in a room. Let's call them **Alison**, **Catherine**, **Emma**, **Gary**, **John**, **Matthew** and **Thomas**. How many ways are there of choosing different groupings of 3 of them? One way would be **A, C, E**; another would be **A, C, T**. Mathematicians find it useful to write $C(n,r)$ to stand for the number in the nth row, in the rth position (counting from $r = 0$) of Pascal's triangle. The answer to our question is $C(7,3)$. The number in the 7th row of the triangle, in the 3rd position, is 35. If we choose one group of 3 we have automatically selected an 'unchosen' group of 4 people. This accounts for the fact that $C(7,4) = 35$ too. In general, $C(n,r) = C(n, n - r)$ which follows from the mirror symmetry of Pascal's triangle.

0s and 1s In Pascal's triangle, we see that the inner numbers form a pattern depending on whether they are even or odd. If we substitute 1 for the odd numbers and 0 for the even numbers we get a representation which is the same pattern as the remarkable fractal known as the Sierpiński gasket (see page 102).

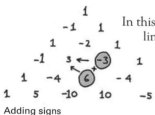

The Serpiński gasket

Adding signs We can write down the Pascal triangle that corresponds to the powers of $(-1 + x)$, namely $(-1 + x)^n$.

In this case the triangle is not completely symmetric about the vertical line, and instead of the rows adding to powers of 2, they add up to zero. However it is the diagonals which are interesting here. The southwestern diagonal $1, -1, 1, -1, 1, -1, 1, -1, \ldots$ are the coefficients of the expansion

Adding signs

$$(1 + x)^{-1} = 1 - x + x^2 - x^3 + x^4 - x^5 + x^6 - x^7 + \ldots$$

while the terms in the next diagonal along are the coefficients of the expansion

$$(1 + x)^{-2} = 1 - 2x + 3x^2 - 4x^3 + 5x^4 - 6x^5 + 7x^6 - 8x^7 + \cdots$$

The Leibniz harmonic triangle

The German polymath Gottfried Leibniz discovered a remarkable set of numbers in the form of a triangle. The Leibniz numbers have a symmetry relation about the vertical line. But unlike Pascal's triangle, the number in one row is obtained by adding the two numbers *below* it. For example $1/30 + 1/20 = 1/12$. To construct this triangle we can progress from the top and move from left to right by subtraction: we know $1/12$ and $1/30$ and so $1/12 - 1/30 = 1/20$, the number next to $1/30$. You might have spotted that the outside diagonal is the famous harmonic series

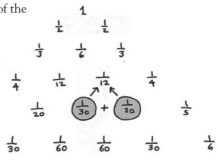

The Leibniz harmonic triangle

$$1 + \frac{1}{2} + \frac{1}{3} + \frac{1}{4} + \frac{1}{5} + \frac{1}{6} + \frac{1}{7} + \cdots$$

but the second diagonal is what is known as the Leibnizian series

$$\frac{1}{1\times2} + \frac{1}{2\times3} + \cdots + \frac{1}{n \times (n + 1)}$$

which by some clever manipulation turns out to equal $n/(n + 1)$. Just as we did before, we can write these Leibnizian numbers as $B(n,r)$ to stand for the nth number in the rth row. They are related to the ordinary Pascal numbers $C(n,r)$ by the formula:

$$B(n,r) \times C(n,r) = \frac{1}{n+1}$$

In the words of the old song, 'the knee bone's connected to the thigh bone, and the thigh bone's connected to the hip bone'. So it is with Pascal's triangle and its intimate connections with so many parts of mathematics – modern geometry, combinatorics and algebra to name but three. More than this it is an exemplar of the mathematical trade – the constant search for pattern and harmony which reinforces our understanding of the subject itself.

the condensed idea
The number fountain

14 Algebra

Algebra gives us a distinctive way of solving problems, a deductive method with a twist. That twist is 'backwards thinking'. For a moment consider the problem of taking the number 25, adding 17 to it, and getting 42. This is forwards thinking. We are given the numbers and we just add them together. But instead suppose we were given the answer 42, and asked a different question? We now want the number which when added to 25 gives us 42. This is where backwards thinking comes in. We want the value of x which solves the equation $25 + x = 42$ and we subtract 25 from 42 to give it to us.

Word problems which are meant to be solved by algebra have been given to schoolchildren for centuries:

> *My niece Michelle is 6 years of age, and I am 40.*
> *When will I be three times as old as her?*

We could find this by a trial and error method but algebra is more economical. In x years from now Michelle will be $6 + x$ years and I will be $40 + x$. I will be three times older than her when

$$3 \times (6 + x) = 40 + x$$

Multiply out the left-hand side of the equation and you'll get $18 + 3x = 40 + x$, and by moving all the xs over to one side of the equation and the numbers to the other, we find that $2x = 22$ which means that $x = 11$. When I am 51 Michelle will be 17 years old. Magic!

What if we wanted to know when I will be *twice* as old as her? We can use the same approach, this time solving

timeline
1950BC

The Babylonians work with
quadratic equations

AD**250**

Diophantus of Alexandria
publishes *Arithmetica*

$$2 \times (6 + x) = 40 + x$$

to get $x = 28$. She will be 34 when I am 68. All the equations above are of the simplest type – they are called 'linear' equations. They have no terms like x^2 or x^3, which make equations more difficult to solve. Equations with terms like x^2 are called 'quadratic' and those with terms like x^3 are called 'cubic' equations. In past times, x^2 was represented as a square and because a square has four sides the term quadratic was used; x^3 was represented by a cube.

Mathematics underwent a big change when it passed from the science of arithmetic to the science of symbols or algebra. To progress from numbers to letters is a mental jump but the effort is worthwhile.

Origins Algebra was a significant element in the work of Islamic scholars in the ninth century. Al-Khwarizmi wrote a mathematical textbook which contained the Arabic word *al-jabr*. Dealing with practical problems in terms of linear and quadratic equations, al-Khwarizmi's 'science of equations' gave us the word 'algebra'. Still later Omar Khayyam is famed for writing the *Rubaiyat* and the immortal lines (in translation)

> *A Jug of Wine, a Loaf of Bread – and Thou*
> *Beside me singing in the Wilderness*

but in 1070, aged 22, he wrote a book on algebra in which he investigated the solution of cubic equations.

Girolamo Cardano's great work on mathematics, published in 1545, was a watershed in the theory of equations for it contained a wealth of results on the cubic equation and the quartic equation – those involving a term of the kind x^4. This flurry of research showed that the quadratic, cubic and quartic equations could all be solved by formulae involving only the operations $+$, $-$, $\times$, $\div$, $\sqrt[q]{}$ (the last operation means the qth root). For example, the

The Italian connection

The theory of cubic equations was fully developed during the Renaissance. Unfortunately it resulted in an episode when mathematics was not always on its best behaviour. Scipione Del Ferro found the solution to the various specialized forms of the cubic equation and, hearing of it, Niccolò Fontana – dubbed 'Tartaglia' or 'the stammerer' – a teacher from Venice, published his own results on algebra but kept his methods secret. Girolamo Cardano from Milan persuaded Tartaglia to tell him of his methods but was sworn to secrecy. The method leaked out and a feud between the two developed when Tartaglia discovered his work had been published in Cardano's 1545 book *Ars Magna*.

quadratic equation $ax^2 + bx + c = 0$ can be solved using the formula:

$$x = \frac{-b \pm \sqrt{b^2 - 4ac}}{2a}$$

If you want to solve the equation $x^2 - 3x + 2 = 0$ all you do is feed the values $a = 1$, $b = -3$ and $c = 2$ into the formula.

The formulae for solving the cubic and quartic equations are long and unwieldy but they certainly exist. What puzzled mathematicians was that they could not produce a formula which was generally applicable to equations involving x^5, the 'quintic' equations. What was so special about the power of five?

In 1826, the short-lived Niels Abel came up with a remarkable answer to this quintic equation conumdrum. He actually proved a negative concept, nearly always a more difficult task than proving that something can be done. Abel proved there could not be a formula for solving all quintic equations, and concluded that any further search for this particular holy grail would be futile. Abel convinced the top rung of mathematicians, but news took a long time to filter through to the wider mathematical world. Some mathematicians refused to accept the result, and well into the 19th century people were still publishing work which claimed to have found the non-existent formula.

The modern world For 500 years algebra meant 'the theory of equations' but developments took a new turn in the 19th century. People realized that symbols in algebra could represent more than just numbers – they could represent 'propositions' and so algebra could be related to the study of logic. They could even represent higher-dimensional objects such as those found in matrix algebra (see page 156). And, as many non-mathematicians have long suspected, they could even represent nothing at all and just be symbols moved about according to certain (formal) rules.

A significant event in modern algebra occurred in 1843 when the Irishman William Rowan Hamilton discovered the quaternions. Hamilton was seeking a system of symbols that would extend two-dimensional complex numbers to higher dimensions. For many years he tried three-dimensional symbols, but no satisfactory system resulted. When he came down for breakfast each morning his sons would ask him, 'Well, Papa, can you *multiply* triplets?' and he was bound to answer that he could only add and subtract them.

Success came rather unexpectedly. The three-dimensional quest was a dead end – he should have gone for four-dimensional symbols. This flash of inspiration came to him as he walked with his wife along the Royal Canal to Dublin. He was ecstatic about the sensation of discovery. Without hesitation, the 38-year-old vandal, Astronomer Royal of Ireland and Knight of the Realm, carved the defining relations into the stone on Brougham Bridge – a spot that is acknowledged today by a plaque. With the date scored into his mind, the subject became Hamilton's obsession. He lectured on it year after year and published two heavyweight books on his 'westward floating, mystic dream of four'.

One peculiarity of quarterions is that when they are multiplied together, the order in which this is done is vitally important, contrary to the rules of ordinary arithmetic. In 1844 the German linguist and mathematician Hermann Grassmann published another algebraic system with rather less drama. Ignored at the time, it has turned out to be far reaching. Today both quaternions and Grassmann's algebra have applications in geometry, physics and computer graphics.

The abstract In the 20th century the dominant paradigm of algebra was the axiomatic method. This had been used as a basis for geometry by Euclid but it wasn't applied to algebra until comparatively recently.

Emmy Noether was the champion of the abstract method. In this modern algebra, the pervading idea is the study of structure where individual examples are subservient to the general abstract notion. If individual examples have the same structure but perhaps different notation they are called isomorphic.

The most fundamental algebraic structure is a group and this is defined by a list of axioms (see page 155). There are structures with fewer axioms (such as groupoids, semi-groups and quasi-groups) and structures with more axioms (like rings, skew-fields, integral domains and fields). All these new words were imported into mathematics in the early 20th century as algebra transformed itself into an abstract science known as 'modern algebra'.

the condensed idea
Solving for the unknown

15 Euclid's algorithm

Al-Khwarizmi gave us the word 'algebra', but it was his ninth-century book on arithmetic that gave us the word 'algorithm'. Pronounced 'Al Gore rhythm' it is a concept useful to mathematicians and computer scientists alike. But what is one? If we can answer this we are on the way to understanding Euclid's division algorithm.

Firstly, an algorithm is a routine. It is a list of instructions such as 'you do this and then you do that'. We can see why computers like algorithms because they are very good at following instructions and never wander off track. Some mathematicians think algorithms are boring because they are repetitious, but to write an algorithm and then translate it into hundreds of lines of computer code containing mathematical instructions is no mean feat. There is a considerable risk of it all going horribly wrong. Writing an algorithm is a creative challenge. There are often several methods available to do the same task and the best one must be chosen. Some algorithms may not be 'fit for purpose' and some may be downright inefficient because they meander. Some may be quick but produce the wrong answer. It's a bit like cooking. There must be hundreds of recipes (algorithms) for cooking roast turkey with stuffing. We certainly don't want a poor algorithm for doing this on the one day of the year when it matters. So we have the ingredients and we have the instructions. The start of the (abbreviated) recipe might go something like this:

- Fill the turkey cavity with stuffing
- Rub the outside skin of the turkey with butter
- Season with salt, pepper and paprika
- Roast at 335 degrees for 3½ hours
- Let the cooked turkey rest for ½ hour

timeline

c.300BC	**c.AD300**
Euclid's algorithm is published in Book 7 of *Elements*	Sun Tzu discovers the Chinese remainder theorem

All we have to do is carry out the algorithm in sequential steps one after the other. The only thing missing in this recipe, usually present in a mathematical algorithm, is a loop, a tool to deal with recursion. Hopefully we won't have to cook the turkey more than once.

In mathematics we have ingredients too – these are the numbers. Euclid's division algorithm is designed to calculate the greatest common divisor (written *gcd*). The *gcd* of two whole numbers is the greatest number that divides into both of them. As our example ingredients, we'll choose the two numbers 18 and 84.

The greatest common divisor The *gcd* in our example is the largest number that exactly divides both 18 and 84. The number 2 divides both 18, and 84, but so does the number 3. So 6 will also divide both numbers. Is there a larger number that will divide them? We could try 9 or 18. On checking, these candidates do not divide 84 so 6 *is* the largest number that divides both. We can conclude that 6 is the *gcd* of 18 and 84, writing this as *gcd*(18, 84) = 6.

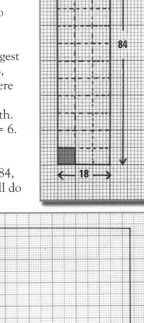

The *gcd* can be interpreted in terms of kitchen tiling. It is the side of the largest square tile that will tile a rectangular wall of breadth 18 and length 84, with no cutting of tiles allowed. In this case, we can see that a 6 × 6 tile will do the trick.

The greatest common divisor is also known as the 'highest common factor' or 'highest common divisor'. There is also a related concept, the least common multiple (*lcm*). The *lcm* of 18 and 84 is the smallest number divisible by both 18 and 84. The link between the *lcm* and *gcd* is highlighted by the fact that the *lcm* of two numbers multiplied by their *gcd* is equal to the multiplication of the two numbers themselves. Here *lcm*(18, 84) = 252 and we can check that 6 × 252 = 1512 = 18 × 84.

Geometrically, the *lcm* is the length of the side of the smallest square that can be tiled by 18 × 84 rectangular tiles. Because *lcm*(a, b) = ab ÷ gcd(a, b), we're going to concentrate on finding the *gcd*. We have already calculated gcd(18, 84) = 6 but to do it we needed to know the divisors of both 18 and 84.

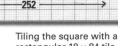

Tiling the square with a rectangular 18 × 84 tile

1202

izmi gives the word
'ı' to mathematics

Fibonacci publishes work on
congruences in *Liber Abaci*

1970s

The Chinese remainder
theorem is applied to
message encryption

Recapping, we first broke both numbers into their factors: $18 = 2 \times 3 \times 3$ and $84 = 2 \times 2 \times 3 \times 7$. Then, comparing them, the number 2 is common to both and is the highest power of 2 which will divide both. Likewise 3 is common and is the highest power dividing both, but though 7 divides 84 it does not divide 18 so it cannot enter into the *gcd* as a factor. We concluded: $2 \times 3 = 6$ is the largest number that divides both. Can this juggling of factors be avoided? Imagine the calculations if we wanted to find *gcd*(17640, 54054). We'd first have to factorize both these numbers, and that would be only the start. There must be an easier way.

The algorithm There is a better way. Euclid's algorithm is given in *Elements*, Book 7, Proposition 2 (in translation): 'Given two numbers not prime to one another, to find their greatest common measure.'

The algorithm Euclid gives is beautifully efficient and effectively replaces the effort of finding factors by simple subtraction. Let's see how it works.

The object is to calculate $d = gcd(18, 84)$. We start by dividing 18 into 84. It does not divide exactly but goes 4 times with 12 (the remainder) left over:

$$84 = \mathbf{4} \times 18 + 12$$

Since d must divide 84 and 18, it must divide the remainder 12. Therefore $d = gcd(12, 18)$. So we can now repeat the process and divide 18 by 12:

$$18 = \mathbf{1} \times 12 + 6$$

to get remainder 6, so $d = gcd(6, 12)$. Dividing 6 into 12 we have remainder 0 so that $d = gcd(0, 6)$. 6 is the largest number which will divide both 0 and 6 so this is our answer.

If computing $d = gcd(17640, 54054)$, the successive remainders would be 1134, 630, 504 and 0, giving us $d = 126$.

Uses for the *gcd* The *gcd* can be used in the solution of equations when the solutions must be whole numbers. These are called Diophantine equations, named after the early Greek mathematician Diophantus of Alexandria.

Let's imagine Great Aunt Christine is going for her annual holiday to Barbados. She sends her butler John down to the airport with her collection of suitcases, each of which weighs either 18 or 84 kilograms, and is informed that the total weight checked-in is 652 kilograms. When he arrives back in Belgravia, John's nine-year-old son James pipes up 'that can't be right, because the *gcd* 6 doesn't divide into 652'. James suggests that the correct total weight might actually be 642 kilograms.

James knows that there is a solution in whole numbers to the equation $18x + 84y = c$ if and only if the *gcd* 6 divides the number c. It does not for $c = 652$ but it does for 642. James does not even need to know how many suitcases x, y of either weight Aunt Christine intends to take to Barbados.

The Chinese remainder theorem When the *gcd* of two numbers is 1 we say they are 'relatively prime'. They don't have to be prime themselves but just have to be prime to each other, for example $gcd(6, 35) = 1$, even though neither 6 nor 35 is prime. We shall need this for the Chinese remainder theorem.

Let's look at another problem: Angus does not know how many bottles of wine he has but, when he pairs them up, there is 1 left over. When he puts them in rows of five in his wine rack there are 3 left over. How many bottles does he have? We know that on division by 2 we get remainder 1 and on division by 5 we get remainder 3. The first condition allows us to rule out all the even numbers. Running along the odd numbers we quickly find that 13 fits the bill (we can safely assume Angus has more than 3 bottles, a number which also satisfies the conditions). But there are other numbers which would also be correct – in fact a whole sequence starting 13, 23, 33, 43, 53, 63, 73, 83, . . .

Let's now add another condition, that the number must give remainder 3 on division by 7 (the bottles arrived in packs of 7 bottles with 3 spares). Running along the sequence 13, 23, 33, 43, 53, 63, . . . to take account of this, we find that 73 fits the bill, but notice that 143 does too, as does 213 and any number found by adding multiples of 70 to these numbers.

In mathematical terms, we have found solutions guaranteed by the Chinese remainder theorem, which also says that any two solutions differ by a multiple of $2 \times 5 \times 7 = 70$. If Angus has between 150 and 250 bottles then the theorem nails the solution down to 213 bottles. Not bad for a theorem discovered in the third century.

The condensed idea
A route to the greatest

16 Logic

'If there are fewer cars on the roads the pollution will be acceptable. Either we have fewer cars on the road or there should be road pricing, or both. If there is road pricing the summer will be unbearably hot. The summer is actually turning out to be quite cool. The conclusion is inescapable: pollution is acceptable.'

Is this argument from the leader of a daily newspaper 'valid' or is it illogical? We are not interested in whether it makes sense as a policy for road traffic or whether it makes good journalism. We are only interested in its validity as a rational argument. Logic can help us decide this question – for it concerns the rigorous checking of reasoning.

Two premises and a conclusion As it stands the newspaper passage is quite complicated. Let's look at some simpler arguments first, going all the way back to the Greek philosopher Aristotle of Stagira who is regarded as the founder of the science of logic. His approach was based on the different forms of the syllogism, a style of argument based on three statements: two premises and a conclusion. An example is

All spaniels are dogs
All dogs are animals
_____
All spaniels are animals

Above the line we have the premises, and below it, the conclusion. In this example, the conclusion has a certain inevitability about it whatever meaning we attach to the words 'spaniels', 'dogs' and 'animals'. The same syllogism, but using different words is

timeline

c.335BC
Aristotle formalizes the logic of the syllogism

AD1847
Boole publishes *The Mathematic Analysis of Logic*

All apples are oranges
All oranges are bananas
──────────────────────
All apples are bananas

In this case, the individual statements are plainly nonsensical if we are using the usual connotations of the words. Yet both instances of the syllogism have the same structure and it is the structure which makes this syllogism valid. It is simply not possible to find an instance of As, Bs and Cs with this structure where the premises are true but the conclusion is false. This is what makes a valid argument useful.

All As are Bs
All Bs are Cs
────────────
All As are Cs

A valid argument

A variety of syllogisms are possible if we vary the quantifiers such as 'All', 'Some' and 'No' (as in No As are Bs). For example, another might be

Some As are Bs
Some Bs are Cs
──────────────
Some As are Cs

Is this a valid argument? Does it apply to *all* cases of As, Bs and Cs, or is there a counterexample lurking, an instance where the premises are true but the conclusion false? What about making A spaniels, B brown objects, and C tables? Is the following instance convincing?

Some spaniels are brown
Some brown objects are tables
─────────────────────────────
Some spaniels are tables

Our counterexample shows that this syllogism is *not* valid. There were so many different types of syllogism that medieval scholars invented mnemonics to help remember them. Our first example was known as B<u>A</u>RB<u>A</u>R<u>A</u> because it contains three uses of 'All'. These methods of analysing arguments lasted for more than 2000 years and held an important place in undergraduate studies in medieval universities. Aristotle's logic – his theory of the syllogism – was thought to be a perfect science well into the 19th century.

1910
Russell and Whitehead attempt to reduce mathematics to logic

1965
Lofti Zadeh develops fuzzy logic

1987
The underground train system in Japan is based on fuzzy logic

a	b	a ∨ b
T	T	T
T	F	T
F	T	T
F	F	F

Or truth table

a	b	a ∧ b
T	T	T
T	F	F
F	T	F
F	F	F

And truth table

a	¬ a
T	F
F	T

Not truth table

a	b	a → b
T	T	T
T	F	F
F	T	T
F	F	T

Implies truth table

Propositional logic Another type of logic goes further than syllogisms. It deals with propositions or simple statements and the combination of them. To analyse the newspaper leader we'll need some knowledge of this 'propositional logic'. It used to be called the 'algebra of logic', which gives us a clue about its structure, since George Boole realized that it could be treated as a new sort of algebra. In the 1840s there was a great deal of work done in logic by such mathematicians as Boole and Augustus De Morgan.

Let's try it out and consider a proposition **a**, where **a** stands for 'Freddy is a spaniel'. The proposition **a** may be True or False. If I am thinking of my dog named Freddy who is indeed a spaniel then the statement is true (**T**) but if I am thinking that this statement is being applied to my cousin whose name is also Freddy then the statement is false (**F**). The truth or falsity of a proposition depends on its reference.

If we have another proposition **b** such as 'Ethel is a cat' then we can combine these two propositions in several ways. One combination is written **a** ∨ **b**. The connective ∨ corresponds to 'or' but its use in logic is slightly different from 'or' in everyday language. In logic, **a** ∨ **b** is true if *either* 'Freddy is a spaniel' is true or 'Ethel is a cat' is true, *or* if both are true, and it is only false when *both* **a** and **b** are false. This conjunction of propositions can be summarized in a truth table.

We can also combine propositions using 'and', written as **a** ∧ **b**, and 'not', written as ¬**a.** The algebra of logic becomes clear when we combine these propositions using a mixture of the connectives with **a**, **b** and **c** like **a** ∧(**b** ∨ **c**). We can obtain an equation we call an identity:

$$a \wedge (b \vee c) \equiv (a \wedge b) \vee (a \wedge c)$$

The symbol ≡ means equivalence between logical statements where both sides of the equivalence have the same truth table. There is a parallel between the algebra of logic and ordinary algebra because the symbols ∧ and ∨ act similarly to × and + in ordinary algebra, where we have $x \times (y + z) = (x \times y) + (x \times z)$. However, the parallel is not exact and there are exceptions.

Other logical connectives may be defined in terms of these basic ones. A useful one is the 'implication' connective **a**→**b** which is defined to be equivalent to ¬**a** ∨ **b** and has the truth table shown.

Now if we look again at the newspaper leader, we can write it in symbolic form to give the argument in the margin:

C = fewer Cars on the roads
P = Pollution will be acceptable
S = there is a road pricing Scheme
H = summer will be unbearably Hot

$$C \rightarrow P$$
$$C \lor S$$
$$S \rightarrow H$$
$$\neg H$$
$$\overline{}$$
$$P$$

Is the argument valid or not? Let's assume the conclusion **P** is false, but that *all* the premises are true. If we can show this forces a contradiction, it means the argument must be valid. It will then be impossible to have the premises true but the conclusion false. If **P** is false, then from the first premise **C → P**, **C** must be false. As **C ∨ S** is true, the fact that **C** is false means that **S** is true. From the third premise **S → H** this means that **H** is true. That is, **¬H** is false. This contradicts the fact that **¬H**, the last premise, was assumed to be true. The content of the statements in the newspaper leader may still be disputed, but the structure of the argument is valid.

Other logics Gottlob Frege, C.S. Peirce, and Ernst Schröder introduced quantification to propositional logic and constructed a 'first-order predicate logic' (because it is predicated on variables). This uses the universal quantifier, ∀, to mean 'for all', and the existential quantifier, ∃, to mean 'there exists'.

∨	or
∧	and
¬	not
→	implies
∀	for all
∃	there exists

Another new development in logic is the idea of fuzzy logic. This suggests confused thinking, but it is really about a widening of the traditional boundaries of logic. Traditional logic is based on collections or sets. So we had the set of spaniels, the set of dogs, and the set of brown objects. We are sure what is included in the set and what is not in the set. If we meet a pure bred 'Rhodesian ridgeback' in the park we are pretty sure it is not a member of the set of spaniels.

Fuzzy set theory deals with what appear to be imprecisely defined sets. What if we had the set of heavy spaniels. How heavy does a spaniel have to be to be included in the set? With fuzzy sets there is a *gradation* of membership and the boundary as to what is in and what is out is left fuzzy. Mathematics allows us to be precise about fuzziness. Logic is far from being a dry subject. It has moved on from Aristotle and is now an active area of modern research and application.

the condensed idea
The clear line of reason

17 Proof

Mathematicians attempt to justify their claims by proofs. The quest for cast iron rational arguments is the driving force of pure mathematics. Chains of correct deduction from what is known or assumed, lead the mathematician to a conclusion which then enters the established mathematical storehouse.

Proofs are not arrived at easily – they often come at the end of a great deal of exploration and false trails. The struggle to provide them occupies the centre ground of the mathematician's life. A successful proof carries the mathematician's stamp of authenticity, separating the established theorem from the conjecture, bright idea or first guess.

Qualities looked for in a proof are rigour, transparency and, not least, elegance. To this add insight. A good proof is 'one that makes us wiser' – but it is also better to have some proof than no proof at all. Progression on the basis of unproven facts carries the danger that theories may be built on the mathematical equivalent of sand.

Not that a proof lasts forever, for it may have to be revised in the light of developments in the concepts it relates to.

What is a proof? When you read or hear about a mathematical result do you believe it? What would make you believe it? One answer would be a logically sound argument that progresses from ideas you accept to the statement you are wondering about. That would be what mathematicians call a proof, in its usual form a mixture of everyday language and strict logic. Depending on the quality of the proof you are either convinced or remain sceptical.

The main kinds of proof employed in mathematics are: the method of the counterexample; the direct method; the indirect method; and the method of mathematical induction.

timeline

c.300 BC	**AD 1637**
Euclid's *Elements* provides the model for mathematical proof	Descartes promotes mathematic rigour as a model in his *Discours on Method*

The counterexample Let's start by being sceptical – this is a method of proving a statement is incorrect. We'll take a specific statement as an example. Suppose you hear a claim that any number multiplied by itself results in an even number. Do you believe this? Before jumping in with an answer we should try a few examples. If we have a number, say 6, and multiply it by itself to get $6 \times 6 = 36$ we find that indeed 36 is an even number. But one swallow does not make a summer. The claim was for *any* number, and there are an infinity of these. To get a feel for the problem we should try some more examples. Trying 9, say, we find that $9 \times 9 = 81$. But 81 is an odd number. This means that the statement that *all* numbers when multiplied by themselves give an even number is false. Such an example runs counter to the original claim and is called a counterexample. A counterexample to the claim that 'all swans are white', would be to see one black swan. Part of the fun of mathematics is seeking out a counterexample to shoot down a would-be theorem.

If we fail to find a counterexample we might feel that the statement is correct. Then the mathematician has to play a different game. A proof has to be constructed and the most straightforward kind is the direct method of proof.

The direct method In the direct method we march forward with logical argument from what is already established, or has been assumed, to the conclusion. If we can do this we have a theorem. We cannot prove that multiplying any number by itself results in an even number because we have already disproved it. But we may be able to salvage something. The difference between our first example, 6, and the counterexample, 9, is that the first number is even and the counterexample is odd. Changing the hypothesis is something we can do. Our new statement is: if we multiply an *even* number by itself the result is an even number.

First we try some other numerical examples and we find this statement verified every time and we just cannot find a counterexample. Changing tack we try to prove it, but how can we start? We could begin with a general even number n, but as this looks a bit abstract we'll see how a proof might go by looking at a concrete number, say 6. As you know, an even number is one which is a multiple of 2, that is $6 = 2 \times 3$. As $6 \times 6 = 6 + 6 + 6 + 6 + 6 + 6$ or, written another way, $6 \times 6 = 2 \times 3 + 2 \times 3 + 2 \times 3 + 2 \times 3 + 2 \times 3 + 2 \times 3$ or, rewriting using brackets,

838

Morgan introduces the term 'mathematical induction'

1967

Bishop proves results exclusively by constructive methods

1976

Imre Lakatos publishes the influential *Proofs and Refutations*

$$6 \times 6 = 2 \times (3 + 3 + 3 + 3 + 3 + 3)$$

This means 6×6 is a multiple of 2 and, as such, is an even number. But in this argument there is nothing which is particular to 6, and we could have started with $n = 2 \times k$ to obtain

$$n \times n = 2 \times (k + k + \ldots + k)$$

and conclude that $n \times n$ is even. Our proof is now complete. In translating Euclid's *Elements*, latter-day mathematicians wrote 'QED' at the end of a proof to say job done – it's an abbreviation for the Latin *quod erat demonstrandum* (which was to be demonstrated). Nowadays they use a filled-in square ■. This is called a halmos after Paul Halmos who introduced it.

The indirect method In this method we pretend the conclusion is false and by a logical argument demonstrate that this contradicts the hypothesis. Let's prove the previous result by this method.

Our hypothesis is that n is even and we'll pretend $n \times n$ is odd. We can write $n \times n = n + n + \ldots + n$ and there are n of these. This means n cannot be even (because if it were $n \times n$ would be even). Thus n is odd, which contradicts the hypothesis. ■

This is actually a mild form of the indirect method. The full-strength indirect method is known as the method of *reductio ad absurdum* (reduction to the absurd), and was much loved by the Greeks. In the academy in Athens, Socrates and Plato loved to prove a debating point by wrapping up their opponents in a mesh of contradiction and out of it would be the point they were trying to prove. The classical proof that the square root of 2 is an irrational number is one of this form where we start off by assuming the square root of 2 is a rational number and deriving a contradiction to this assumption.

The method of mathematical induction Mathematical induction is powerful way of demonstrating that a sequence of statements P_1, $P_2, P_3, \ldots$ are all true. This was recognized by Augustus De Morgan in the 1830s who formalized what had been known for hundreds of years. This specific technique (not to be confused with scientific induction) is widely used to prove statements involving *whole* numbers. It is especially useful in graph theory, number theory, and computer science generally. As a practical example, think of the problem of adding up the odd numbers. For instance, the addition of the first three odd numbers $1 + 3 + 5$ is 9 while the sum of first four $1 + 3 + 5 + 7$ is 16. Now 9 is $3 \times 3 = 3^2$ and 16 is $4 \times 4 = 4^2$, so could it be that the addition of the first n odd numbers is equal to n^2? If we try a randomly chosen value of n, say $n = 7$, we indeed find that the sum of the first seven is $1 + 3 + 5 + 7 + 9 + 11 + 13 = 49$ which is 7^2. But is this pattern followed for *all* values of n? How can we be sure? We have a problem, because we cannot hope to check an infinite number of cases individually.

This is where mathematical induction steps in. Informally it is the domino method of proof. This metaphor applies to a row of dominos standing on their ends. If one domino falls it will knock the next one down. This is clear. All we need to make them *all* fall is the first one to fall. We can apply this thinking to the odd numbers problem. The statement P_n says that the sum of the first n odd numbers adds up to n^2. Mathematical induction sets up a chain reaction whereby $P_1, P_2, P_3, \ldots$ will *all* be true. The statement P_1 is trivially true because $1 = 1^2$. Next, P_2 is true because $1 + 3 = 1^2 + 3 = 2^2$, P_3 is true because $1 + 3 + 5 = 2^2 + 5 = 3^2$ and P_4 is true because $1 + 3 + 5 + 7 = 3^2 + 7 = 4^2$. We use the result at one stage to hop to the next one. This process can be formalized to frame the method of mathematical induction.

Difficulties with proof Proofs come in all sorts of styles and sizes. Some are short and snappy, particularly those found in the text books. Some others detailing the latest research have taken up the whole issue of journals and amount to thousands of pages. Very few people will have a grasp of the whole argument in these cases.

There are also foundational issues. For instance, a small number of mathematicians are unhappy with the reductio ad absurdam method of indirect proof where it applies to existence. If the assumption that a solution of an equation does not exist leads to a contradiction, is this enough to prove that a solution does exist? Opponents of this proof method would claim the logic is merely sleight of hand and doesn't tell us how to actually construct a concrete solution. They are called 'Constructivists' (of varying shades) who say the proof method fails to provide 'numerical meaning'. They pour scorn on the classical mathematician who regards the reductio method as an essential weapon in the mathematical armoury. On the other hand the more traditional mathematician would say that outlawing this type of argument means working with one hand tied behind your back and, furthermore, denying so many results proved by this indirect method leaves the tapestry of mathematics looking rather threadbare.

the condensed idea
Signed and sealed

18 Sets

Nicholas Bourbaki was a pseudonym for a self-selected group of French academics who wanted to rewrite mathematics from the bottom up in 'the right way'. Their bold claim was that everything should be based on the theory of sets. The axiomatic method was central and the books they put out were written in the rigorous style of 'definition, theorem and proof'. This was also the thrust of the modern mathematics movement of the 1960s.

Georg Cantor created set theory out of his desire to put the theory of real numbers on a sound basis. Despite initial prejudice and criticism, set theory was well established as a branch of mathematics by the turn of the 20th century.

What are sets? A set may be regarded as a collection of objects. This is informal but gives us the main idea. The objects themselves are called 'elements' or 'members' of the set. If we write a set A which has a member a, we may write $a \in A$, as did Cantor. An example is A = {1, 2, 3, 4, 5} and we can write $1 \in A$ for membership, and $6 \notin A$ for non-membership.

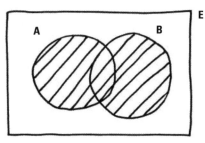

The union of A and B

Sets can be combined in two important ways. If A and B are two sets then the set consisting of elements which are members of A *or* B (or both) is called the 'union' of the two sets. Mathematicians write this as $A \cup B$. It can also be described by a Venn diagram, named after the Victorian logician the Rev. John Venn. Euler used diagrams like these even earlier.

The set $A \cap B$ consists of elements which are members of A *and* B and is called the 'intersection' of the two sets.

AD 1872	1881
Cantor takes a tentative step in the creation of set theory	Venn popularizes 'Venn diagrams' for sets

If $A = \{1, 2, 3, 4, 5\}$ and $B = \{1, 3, 5, 7, 10, 21\}$, the union is $A \cup B = \{1, 2, 3, 4, 5, 7, 10, 21\}$ and the intersection is $A \cap B = \{1, 3, 5\}$. If we regard a set A as part of a universal set E, we can define the complement set $\neg A$ as consisting of those elements in E which are *not* in A.

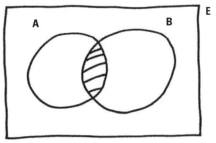

The operations $\cap$ and $\cup$ on sets are analogous to $\times$ and $+$ in algebra. Together with the complement operation $\neg$, there is an 'algebra of sets'. The Indian-born British mathematician Augustus De Morgan, formulated laws to show how all three operations work together. In our modern notation, De Morgan's laws are:

The intersection of A and B

$$\neg(A \cup B) = (\neg A) \cap (\neg B)$$

and

$$\neg(A \cap B) = (\neg A) \cup (\neg B)$$

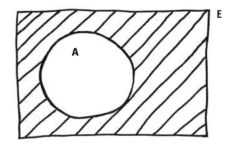

The complement of A

The paradoxes There are no problems dealing with finite sets because we can list their elements, as in $A = \{1, 2, 3, 4, 5\}$, but in Cantor's time, infinite sets were more challenging.

Cantor defined sets as the collection of elements with a specific property. Think of the set $\{11, 12, 13, 14, 15, \ldots\}$, all the whole numbers bigger than 10. Because the set is infinite, we can't write down all its elements, but we can still specify it because of the property that all its members have in common. Following Cantor's lead, we can write the set as $A = \{x: x$ is a whole number $> 10\}$, where the colon stands for 'such that'.

In primitive set theory we could also have a set of abstract things, $A = \{x: x$ is an abstract thing$\}$. In this case A is itself an abstract thing, so it is possible to have $A \in A$. But in allowing this relation, serious problems arise. The British philosopher Bertrand Russell hit upon the idea of a set S which contained *all* things which did *not* contain themselves. In symbols this is $S = \{x: x \notin x\}$.

He then asked the question, 'is $S \in S$?' If the answer is 'Yes' then S must satisfy the defining sentence for S, and so $S \notin S$. On the other hand if the answer is

1931

Gödel proves that any formal axiomatic mathematical system contains undecidable statements

1939

The pseudonym Bourbaki is first used by French mathematicians

1964

Cohen proves the independence of the continuum hypothesis

'No' and $S \notin S$, then S does *not* satisfy the defining relation of $S = \{x: x \notin x\}$ and so $S \in S$. Russell's question ended with this statement, the basis of Russell's paradox,

$$S \in S \text{ if and only if } S \notin S$$

It is similar to the 'barber paradox' where a village barber announces to the locals that he will only shave those who do not shave themselves. The question arises: should the barber shave himself? If he does not shave himself he should. If does shave himself he should not.

It is imperative to avoid such paradoxes, politely called 'antinomies'. For mathematicians it is simply not permissible to have systems that generate contradictions. Russell created a theory of types and only allowed $a \in A$ if a were of a lower type than A, so avoiding expressions such as $S \in S$.

Another way to avoid these antinomies was to formalize the theory of sets. In this approach we don't worry about the nature of sets themselves, but list formal axioms that specify rules for treating them. The Greeks tried something similar with a problem of their own – they didn't have to explain what straight lines were, but only how they should be dealt with.

In the case of set theory, this was the origin of the Zermelo–Fraenkel axioms for set theory which prevented the appearance of sets in their system that were too 'big'. This effectively debarred such dangerous creatures as the set of all sets from appearing.

Gödel's theorem Austrian Mathematician Kurt Gödel dealt a knockout punch to those who wanted to escape from the paradoxes into formal axiomatic systems. In 1931, Gödel proved that even for the simplest of formal systems there were statements whose truth or falsity could not be deduced from within these systems. Informally, there were statements which the axioms of the system could not reach. They were undecidable statements. For this reason Gödel's theorem is paraphrased as 'the incompleteness theorem'. This result applied to the Zermelo–Fraenkel system as well as to other systems.

Cardinal numbers The number of elements of a finite set is easy to count, for example $A = \{1, 2, 3, 4, 5\}$ has 5 elements or we say its 'cardinality' is 5 and write $card(A) = 5$. Loosely speaking, the cardinality measures the 'size' of a set.

According to Cantor's theory of sets, the set of fractions **Q** and the real numbers **R** are very different. The set **Q** can be put in a list but the set **R** cannot (see page 31). Although both sets are infinite, the set **R** has a higher order of infinity than **Q**. Mathematicians denote $card(Q)$ by $\aleph_0$, the Hebrew 'aleph nought' and $card(R) = c$. So this means $\aleph_0 < c$.

The continuum hypothesis Brought to light by Cantor in 1878, the continuum hypothesis says that the next level of infinity after the infinity of **Q** is the infinity of the real numbers c. Put another way, the continuum hypothesis asserted there was no set whose cardinality lay strictly between $\aleph_0$ and c. Cantor struggled with it and though he believed it to be true he could not prove it. To disprove it would amount to finding a subset X of **R** with $\aleph_0 < card(X) < c$ but he could not do this either.

The problem was so important that German mathematician David Hilbert placed it at the head of his famous list of 23 outstanding problems for the next century, presented to the International Mathematical Congress in Paris in 1900.

Gödel emphatically believed the hypothesis to be false, but he did not prove it. He did prove (in 1938) that the hypothesis was compatible with the Zermelo–Fraenkel axioms for set theory. A quarter of a century later, Paul Cohen startled Gödel and the logicians by proving that the continuum hypothesis could not be deduced from the Zermelo–Fraenkel axioms. This is equivalent to showing the axioms and the negation of the hypothesis is consistent. Combined with Gödel's 1938 result, Cohen had shown that the continuum hypothesis was independent of the rest of the axioms for set theory.

This state of affairs is similar in nature to the way the parallel postulate in geometry (see page 108) is independent of Euclid's other axioms. That discovery resulted in a flowering of the non-Euclidean geometries which, amongst other things, made possible the advancement of relativity theory by Einstein. In a similar way, the continuum hypothesis can be accepted or rejected without disturbing the other axioms for set theory. After Cohen's pioneering result a whole new field was created which attracted generations of mathematicians who adopted the techniques he used in proving the independence of the continuum hypothesis.

the condensed idea
Many treated as one

19 Calculus

A calculus is a way of calculating, so mathematicians sometimes talk about the 'calculus of logic', the 'calculus of probability', and so on. But all are agreed there is really only one Calculus, pure and simple, and this is spelled with a capital C.

Calculus is a central plank of mathematics. It would now be rare for a scientist, engineer or a quantitative economist not to have come across Calculus, so wide are its applications. Historically it is associated with Isaac Newton and Gottfried Leibniz who pioneered it in the 17th century. Their similar theories resulted in a priority dispute over who was the discoverer of Calculus. In fact, both men came to their conclusions independently and their methods were quite different.

Since then Calculus has become a huge subject. Each generation bolts on techniques they think should be learned by the younger generation, and these days textbooks run beyond a thousand pages and involve many extras. For all these add-ons, what is absolutely essential is *differentiation* and *integration*, the twin peaks of Calculus as set up by Newton and Leibniz. The words are derived from Leibniz's *differentialis* (taking differences or 'taking apart') and *integralis* (the sum of parts, or 'bringing together').

In technical language, differentiation is concerned with measuring *change* and integration with measuring *area*, but the jewel in the crown of Calculus is the 'star result' that they are two sides of the same coin – differentiation and integration are the inverses of each other. Calculus is really one subject, and you need to know about both sides. No wonder that Gilbert and Sullivan's 'very model of a modern Major General' in *The Pirates of Penzance* proudly proclaimed them both:

With many cheerful facts about the square of the hypotenuse.
I'm very good at integral and differential calculus.

timeline

Differentiation Scientists are fond of conducting 'thought experiments' – Einstein especially liked them. Imagine we are standing on a bridge high above a gorge and are about to let a stone drop. What will happen? The advantage of a thought experiment is that we do not actually have to be there in person. We can also do impossible things like stopping the stone in mid-air or watching it in slow motion over a short time interval.

According to Newton's theory of gravity, the stone will fall. Nothing surprising in that; the stone is attracted to the earth and will fall faster and faster as the hand on our stopwatch ticks on. Another advantage of a thought experiment is that we can ignore complicating factors like air resistance.

What is the stone's speed at a given instant of time, say when the stopwatch reads *exactly* 3 seconds after it has been released? How can we work this out? We can certainly measure *average* speed but our problem is to measure *instantaneous* speed. As it's a thought experiment, why don't we stop the stone in mid-air and then let it move down a short distance by taking a fraction of a second more? If we divide this extra distance by the extra time we will have the average speed over the short time interval. By taking smaller and smaller time intervals the average speed will be closer and closer to the instantaneous speed at the place where we stopped the stone. This limiting process is the basic idea behind Calculus.

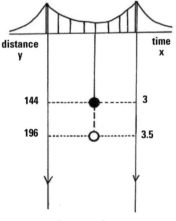

We might be tempted to make the small extra time equal to zero. But in our thought experiment, the stone has not moved at all. It has moved no distance and taken no time to do it! This would give us the average speed 0/0 which the Irish philosopher Bishop Berkeley famously described as the 'ghosts of departed quantities'. This expression cannot be determined – it is actually *meaningless*. By taking this route we are led into a numerical quagmire.

To go further we need some symbols. The exact formula connecting the distance fallen y and the time x taken to reach there was derived by Galileo:

$$y = 16 \times x^2$$

The factor '16' appears because feet and seconds are the chosen measurement units. If we want to know, say, how far the stone has dropped in 3 seconds we simply substitute $x = 3$ into the formula and calculate the answer $y = 16 \times 3^2$ = 144 feet. But how can we calculate the *speed* of the stone at time $x = 3$?

Let's take a further 0.5 of a second and see how far the stone has travelled between 3 and 3.5 seconds. In 3.5 seconds the stone has travelled $y = 16 \times 3.5^2$ = 196 feet, so *between* 3 and 3.5 seconds it has fallen 196 − 144 = 52 feet. Since speed is distance divided by time, the average speed over this time interval is 52/0.5 = 104 feet per second. This will be close to the instantaneous speed at $x = 3$, but you may well say that 0.5 seconds is not a small enough measure. Repeat the argument with a smaller time gap, say 0.05 seconds, and we see that the distance fallen is 148.84 − 144 = 4.84 feet giving an average speed of 4.84/0.05 = 96.8 feet per second. This indeed will be closer to the instantaneous speed of the stone at 3 seconds (when $x = 3$).

We must now take the bull by the horns and address the problem of calculating the average speed of the stone between x seconds and slightly later at $x + h$ seconds. After a little symbol shuffling we find this is

$$16 \times (2x) + 16 \times h$$

As we make h smaller and smaller, like we did in going from 0.5 to 0.05, we see that the first term is unaffected (because it does not involve h) and the second term itself becomes smaller and smaller. We conclude that

u	du/dx
x^2	$2x$
x^3	$3x^2$
x^4	$4x^3$
x^5	$5x^4$
. . .	. . .
x^n	nx^{n-1}

$$v = 16 \times (2x)$$

where v is the instantaneous velocity of the stone at time x. For example, the instantaneous velocity of the stone after 1 second (when $x = 1$) is $16 \times (2 \times 1)$ = 32 feet per second; after 3 seconds it is $16 \times (2 \times 3)$ which gives 96 feet per second.

If we compare Galileo's distance formula $y = 16 \times x^2$ with the velocity formula $v = 16 \times (2x)$ the essential difference is the change x^2 to $2x$. This is the effect of differentiation, passing from $u = x^2$ to the *derivative* $\dot{u} = 2x$. Newton called $\dot{u}$ = $2x$ a 'fluxion' and the variable x a fluent because he thought in terms of flowing quantities. Nowadays we frequently write $u = x^2$ and its *derivative* as $du/dx = 2x$. Originally introduced by Leibniz, this notation's continued use represents the success of the 'd'ism of Leibniz over the dotage of Newton'.

The falling stone was one example, but if we had other expressions that u stood for we could still calculate the derivative, which can be useful in other contexts. There is a pattern in this: the derivative is formed by multiplying by the previous power and subtracting 1 from it to make the new power.

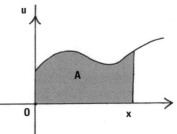

Integration The first application of integration was to measure area. The measurement of the area under a curve is done by dividing it into approximate rectangular strips, each with width dx. By measuring the area of each and adding them up we get the 'sum' and so the total area. The notation S standing for sum was introduced by Leibniz in an elongated form $\int$. The area of each of the rectangular strips is $u\,dx$, so the area A under the curve from 0 to x is

$$A = \int_0^x u\,dx$$

If the curve we're looking at is $u = x^2$, the area is found by drawing narrow rectangular strips under the curve, adding them up to calculate the approximate area, and applying a limiting process to their widths to gain the exact area. This answer gives the area

$$A = x^3/3$$

For different curves (and so other expressions for u) we could still calculate the integral. Like the derivative, there is a regular pattern for the integral of powers of x. The integral is formed by dividing by the 'previous power +1' and adding 1 to it to make the new power.

u	$\int_0^x u\,dx$
x^2	$x^3/3$
x^3	$x^4/4$
x^4	$x^5/5$
x^5	$x^6/6$
$\cdots$	$\cdots$
x^n	$x^{n+1}/(n+1)$

The star result If we differentiate the integral $A = x^3/3$ we actually get the original $u = x^2$. If we integrate the derivative $du/dx = 2x$ we also get the original $u = x^2$. Differentiation is the inverse of integration, an observation known as the Fundamental Theorem of the Calculus and one of the most important theorems in all mathematics.

Without Calculus there would be no satellites in orbit, no economic theory and statistics would be a very different discipline. Wherever change is involved, there we find Calculus.

the condensed idea
Going to the limit

20 Constructions

Proving a negative is often difficult, but some of the greatest triumphs in mathematics do just that. This means proving something cannot be done. Squaring the circle is impossible but how can we prove this?

The Ancient Greeks had four great construction problems:

- trisecting the angle (dividing an angle into three equal smaller angles),
- doubling the cube (building a second cube with twice the volume of the first),
- squaring the circle (creating a square with the same area as a particular circle),
- constructing polygons (building regular shapes with equal sides and angles).

To perform these tasks they only used the bare essentials:

- a straight edge for drawing straight lines (and definitely *not* to measure lengths),
- a pair of compasses for drawing circles.

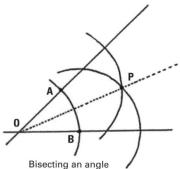

Bisecting an angle

If you like climbing mountains without ropes, oxygen, mobile phones and other paraphernalia, these problems will undoubtedly appeal. Without modern measuring equipment the mathematical techniques needed to prove these results were sophisticated and the classical construction problems of antiquity were only solved in the 19th century using the techniques of modern analysis and abstract algebra.

Trisecting the angle Here is a way to divide an angle into two equal smaller angles or, in other words, bisect it. First place the compass point at O and, with any radius mark off OA and OB. Moving the compass point to A, draw a portion of a circle. Do the same at B. Label the point of intersection of these circles P, and with

Anaxogoras attempts to square the circle while in prison

Mohr shows that all Euclidean constructions can be carried out with compasses alone

the straight edge join O to P. The triangles AOP and BOP are identical in shape and therefore the angles AÔP and BÔP will be equal. The line OP is the required bisector, splitting the angle into two equal angles.

Can we use a sequence of actions like this to split an arbitrary angle into *three* equal angles? This is the angle trisection problem.

If the angle is 90 degrees, a right angle, there is no problem, because the angle of 30 degrees can be constructed. But, if we take the angle of 60 degrees, for instance this angle *cannot* be trisected. We know the answer is 20 degrees but there is no way of constructing this angle using only a straight edge and compasses. So summarizing:

- you can bisect *all* angles *all* the time,
- you can trisect *some* angles *all* the time, but
- you cannot trisect *some* angles at *any* time.

The duplication of the cube is a similar problem known as the Delian problem. The story goes that the natives of Delos in Greece consulted the oracle in the face of a plague they were suffering. They were told to construct a new altar, twice the volume of the existing one.

Imagine the Delian altar began as a three-dimensional cube with all sides equal in length, say a. So they needed to construct another cube of length b with twice its volume. The volume of each is a^3 and b^3 and they are related by $b^3 = 2a^3$ or $b = \sqrt[3]{2} \times a$ where $\sqrt[3]{2}$ is the number multiplied by itself three times that makes 2 (the cube root). If the side of the original cube is $a = 1$ the natives of Delos had to mark off the length $\sqrt[3]{2}$ on a line. Unfortunately for them, this is impossible with a straight edge and compasses no matter how much ingenuity is brought to bear on the would-be construction.

Squaring the circle
This problem is a little different and is the most famous of the construction problems:

To construct a square whose area is equal to the area of a given circle.

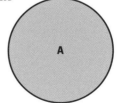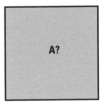

Squaring the circle

801

auss publishes *Discourses on rithmetic* including a section the construction of a regular -gon by ruler and compasses

1837

Wantzel proves that the classical problems of duplicating a cube and trisecting an angle cannot be solved with ruler and compass

1882

Lindemann proves the circle cannot be squared

The phrase 'squaring the circle' is commonly used to express the impossible. The algebraic equation $x^2 - 2 = 0$ has specific solutions $x = \sqrt{2}$ and $x = -\sqrt{2}$. These are irrational numbers (they cannot be written as fractions), but showing the circle cannot be squared amounts to showing that π cannot be a solution of *any* algebraic equation. Irrational numbers with this property are called transcendental numbers because they have a 'higher' irrationality than their irrational cousins like $\sqrt{2}$.

Mathematicians generally believed that π was a transcendental but this 'riddle of the ages'was difficult to prove until Ferdinand von Lindemann used a modification of a technique pioneered by Charles Hermite. Hermite had used it to deal with the lesser problem of proving that the base of natural logarithms, *e*, was transcendental (see page 26).

Following Lindemann's result, we might think that the flow of papers from the indomitable band of 'circle-squarers' would cease. Not a bit of it. Still dancing on the sidelines of mathematics were those reluctant to accept the logic of the proof and some who had never heard of it.

Constructing polygons Euclid posed the problem of how to construct a regular polygon. This is a symmetrical many-sided figure like a square or pentagon, in which sides are all of equal length and where adjacent sides make equal angles with each other. In his famous work the *Elements* (Book 4), Euclid showed how the polygons with 3, 4, 5 and 6 sides could be constructed using only our two basic tools.

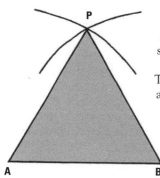

The polygon with 3 sides is what we normally call an equilateral triangle and is particularly straightforward to construct. However long you want your triangle to be, label one point A and another B with the desired distance in between. Place the compass point at A and draw a portion of the circle of radius AB. Repeat this with the compass point at B using the same radius. The intersection point of these two arcs is at P. As AP = AB and BP = AB all three sides of the triangle APB are equal. The actual triangle is completed by joining AB, AP and BP using the straight edge.

Constructing an equilateral triangle

If you think having a straight edge seems rather a luxury, you're not alone – the Dane Georg Mohr thought so too. The equilateral triangle is constructed by finding the point P and for this only the compasses are required – the straight edge was only used to *physically* join the points together. Mohr showed that any construction achievable by straight edge and compasses can be achieved with the compasses alone. The Italian Lorenzo Mascheroni proved the same results 125 years later. A novel feature of his 1797 book *Geometria del Compasso*, dedicated to Napoleon, is that he wrote it in verse.

For the general problem, the polygons with p sides where p is a prime number are especially important. We have already constructed the 3-sided polygon, and Euclid constructed the 5-sided polygon but he could *not* construct the 7-sided polygon (the heptagon). Investigating this problem as a 17 year old, a certain Carl Friederich Gauss proved a negative. He deduced that it is not possible to construct a p-sided polygon for $p = 7$, 11 or 13.

But Gauss also proved a positive, and he concluded that it is possible to construct a 17-sided polygon. Gauss actually went further and proved that a p-sided polygon is constructable if and only if the prime number p is of the form

$$p = 2^{2^n} + 1$$

A prince is born

Carl Friedrich Gauss was so impressed by his result showing a 17-sided polygon could be constructed that he decided to put away his planned study of languages and become a mathematician. The rest is history – and he became known as the 'prince of mathematicians'. The 17-sided polygon is the shape of the base of his memorial at Göttingen, Germany, and is a fitting tribute to his genius.

Numbers of this form are called Fermat numbers. If we evaluate them for $n = 0$, 1, 2, 3 and 4, we find they are the prime numbers $p = 3, 5, 17, 257$ and 65,537, and these correspond to a constructible polygon with p sides.

When we try $n = 5$, the Fermat number is $p = 2^{32} + 1 = 4,294,967,297$. Pierre de Fermat conjectured that they were all prime numbers, but unfortunately this one is not a prime number, because $4,294,967,297 = 641 \times 6,700,417$. If we put $n = 6$ or 7 into the formula the results are huge Fermat numbers but, as with 5, neither is prime.

Are there any other Fermat primes? The accepted wisdom is that there are not, but no one knows for sure.

the condensed idea
Take a straight edge and a pair of compasses . . .

21 Triangles

The most obvious fact about a triangle is that it is a figure with three sides and three angles (tri-angles). Trigonometry is the theory which we use to 'measure the triangle', whether it is the size of the angles, the length of the sides, or the enclosed area. This shape – one of the simplest of all figures – is of enduring interest.

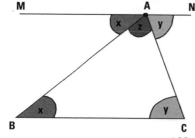

The triangle's tale There is a neat argument to show that the angles in any triangle add up to two right angles or 180 degrees. Through the point or 'vertex' A of any triangle draw a line MAN parallel to the base BC.

The angle $A\hat{B}C$ which we'll call x is equal to the angle $B\hat{A}M$ because they are alternate angles and MN and BC are parallel. The other two alternate angles are equal to y. The angle about the point A is equal to 180 degrees (half of 360 degrees) and this is x + y + z which is the sum of the angles in the triangle. QED as was often written at the end of his proofs. Of course we are assuming the triangle is drawn on a flat surface like this flat piece of paper. The angles of a triangle drawn on a ball (a spherical triangle) do not add up to 180 degrees but that is another story.

Euclid proved many theorems about triangles, always making sure this was done deductively. He showed, for example, that 'in any triangle two sides taken together in any manner are greater than the remaining one'. Nowadays this is called the 'triangle inequality' and is important in abstract mathematics. The Epicureans, with their down-to-earth approach to life, claimed this required no proof, for it was evident even to an ass. If a bale of hay were placed at one vertex and the ass at the other, they argued, the animal would hardly traverse the two sides to satisfy its hunger.

Pythagoras's theorem The greatest triangle theorem of all is Pythagoras's theorem, and is one which features in modern mathematics –

1850BC

The Babylonians know
'Pythagoras's theorem'

AD**1335**

Richard of Wallingford writes a
ground-breaking treatise on
trigonometry

though there is some doubt about Pythagoras being the first to discover it. The best known statement of it is in terms of algebra, $a^2 + b^2 = c^2$ but Euclid refers to actual square shapes: 'In right-angled triangles the square on the side subtending the right angle is equal to the squares on the sides containing the right angle'.

Euclid's proof is Proposition 47, in Book 1 of the *Elements*, a proof which became a point of anxiety for generations of school pupils as they struggled to commit it to memory, or take the consequences. There are several hundred proofs in existence. A favourite is more in the spirit of Bhāskara of the 12th century than a Euclidean proof from 300 BC.

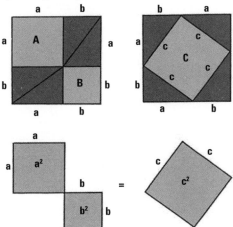

This is a proof 'without words'. In the figure the square with side $a + b$ can be divided in two different ways.

Since the four equal triangles (shaded dark) are common to both squares we can remove them and still have equality of area. If we look at the areas of the remaining shapes, out springs the familiar expression.

$$a^2 + b^2 = c^2$$

The Euler line Hundreds of propositions about triangles are possible. First, let's think about the midpoints of the sides. In any triangle ABC we mark the midpoints D, E, F of its sides. Join B to F and C to D and mark the point where they cross as G. Now join A to E. Does this line also pass through G? It is not obvious that it should necessarily without further reasoning. It fact it does and the point G is called the 'centroid' of the triangle. This is the centre of gravity of the triangle.

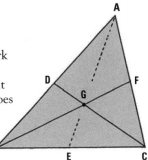

There are literally hundreds of different 'centres' connected with a triangle. Another one is the point H where the altitudes (the lines drawn from a vertex perpendicular to a base – shown as dotted lines in the figure on page 86) meet. This is called the 'orthocentre'. There is also another centre called the 'circumcentre' O where each of the lines (known as 'perpendiculars') at D, E and F meet (not shown). This is the centre of the circle which can be drawn through A, B and C.

1571	1822	1873
François Viète publishes a book on trigonometry and trigonometric tables	Karl Feuerbach describes the nine point circle of a triangle	Brocard produces his exhaustive work on the triangle

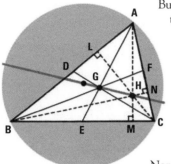

The Euler line

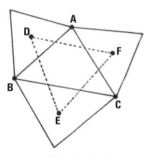

Napoleon's theorem

But more is true. In *any* triangle ABC the centres G, H and O, respectively the centroid, orthocentre, and circumcentre, themselves lie along one line, called the '*Euler line*'. In the case of an equilateral triangle (all sides of equal length) these three points coincide and the resulting point is unambiguously *the* centre of the triangle.

Napoleon's theorem For any triangle ABC, equilateral triangles can be constructed on each side and from their centres a new triangle DEF is constructed. Napoleon's theorem asserts that for *any* triangle ABC, the triangle DEF is an *equilateral* triangle.

Napoleon's theorem appeared in print in an English journal in 1825 a few years after his death on St Helena in 1821. Napoleon's ability in mathematics in school no doubt helped him gain entrance to the artillery school, and later he got to know the leading mathematicians in Paris when he was Emperor. Unfortunately there is no evidence to take us further and 'Napoleon's theorem' is, like many other mathematical results, ascribed to a person who had little to do with its discovery or its proof. Indeed, it is a theorem which is frequently being rediscovered and extended.

The essential data that determines a triangle consists of knowing the length of one side and two angles. By using trigonometry we can measure everything else.

In surveying areas of land in order to draw maps it is quite useful to be a 'flat-earther' and assume triangles to be flat. A network of triangles is established by starting with a base line BC of known length, choosing a distant point A (the triangulation point) and measuring the angles $A\hat{B}C$ and $A\hat{C}B$ by theodolite. By trigonometry everything is known about the triangle ABC and the surveyor moves on, fixes the next triangulation point from the new base line AB or AC and repeats the operation to establish a web of triangles. The method has the advantage of being able to map inhospitable country involving such barriers as marshland, bogs, quicksand and rivers.

It was used as the basis for the Great Trigonometrical Survey of India which began in the 1800s and lasted 40 years. The object was to survey and map along the Great Meridional Arc from Cape Comorin in the south to the Himalayas in the north, a distance of some 1500 miles. To ensure utmost accuracy in measuring angles, Sir George Everest arranged the manufacture of two giant theodolites in London, together weighing one ton and needing teams of a dozen men to transport them. It was vital to get the angles right. Accuracy in measurement was paramount and much talked about but it was the humble triangle which was at the centre of operations. The Victorians had to make do

Building with triangles

The triangle is indispensable in building. Its use and strength relies on the fact that made it indispensable in surveying – a triangle is rigid. You can push a square or rectangle out of shape but not a triangle. The truss used in building is the joining together of triangles, and is seen as the component in roofs. One breakthrough occurred in the building of bridges.

A Warren truss can take a heavy loading compared to its weight. It was patented in 1848 by James Warren and the first bridge designed in this way was constructed at London Bridge Station two years later. The use of equilateral triangles proved more reliable than similar designs based on the isosceles triangles, the triangles where only two sides are required to be equal.

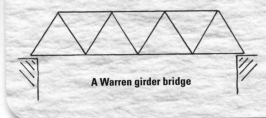

A Warren girder bridge

without GPS though they did have computers – human computers. Once all the lengths in a triangle have been computed, the calculation of area is straightforward. Once again, the triangle is the unit. There are several formulae for the area A of a triangle, but the most remarkable is Heron of Alexandria's formula:

$$A = \sqrt{s \times (s-a) \times (s-b) \times (s-c)}$$

It can be applied to any triangle and we don't even have to know any angles. The symbol s stands for one half of the perimeter of the triangle whose sides are of length a, b and c. For example, if a triangle has sides 13, 14 and 15, the perimeter is $13 + 14 + 15 = 42$, so that $s = 21$. Completing the calculation, $A = \sqrt{(21 \times 8 \times 7 \times 6)} = \sqrt{7056} = 84$. The triangle is a familiar object, whether to children playing with simple shapes or researchers dealing on a day-to-day basis with the triangle inequality in abstract mathematics. Trigonometry is the basis for making calculations about triangles and the sine, cosine and tangent functions are the tools for describing them, enabling us to make accurate calculations for practical applications. The triangle has received much attention but it is surprising that so much is waiting to be discovered about three lines forming such a basic figure.

the condensed idea
Three sides of a story

22 Curves

It's easy to draw a curve. Artists do it all the time; architects lay out a sweep of new buildings in the curve of a crescent, or a modern close. A baseball pitcher throws a curveball. Sportspeople make their way up the pitch in a curve, and when they shoot for goal, the ball follows a curve. But, if we were to ask 'What is a curve?' the answer is not so easy to frame.

Mathematicians have studied curves for centuries and from many vantage points. It began with the Greeks and the curves they studied are now called the 'classical' curves.

Classical curves The first family in the realm of the classical curves are what we call 'conic sections'. Members of this family are the circle, the ellipse, the parabola, and the hyperbola. The conic is formed from the double cone, two ice-cream cones joined together where one is upside down. By slicing through this with a flat plane the curves of intersection will be a circle, an ellipse, a parabola or a hyperbola, depending on the tilt of the slicing plane to the vertical axis of the cone.

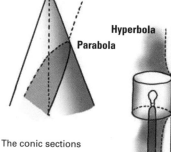

The conic sections

We can think of a conic as the projection of a circle onto a screen. The light rays from the bulb in a cylindrical table lamp form a double light cone where the light will throw out projections of the top and bottom circular rims. The image on the ceiling will be a circle but if we tip the lamp, this circle will become an ellipse. On the other hand the image against the wall will give the curve in two parts, the hyperbola.

The conics can also be described from the way points move in the plane. This is the 'locus' method loved by

timeline

c.300BC	**c.250**BC	**c.225**BC
Euclid defines the conic sections	Archimedes investigates spirals	Apollonius of Perga publishes *Conics*

the Greeks, and unlike the projective definition it involves length. If a point moves so that its distance from *one* fixed point is always the same, we get a circle. If a point moves so that the sum of its distances from *two* fixed points (the foci) is a constant value we get an ellipse (where the two foci are the same, the ellipse becomes a circle). The ellipse was the key to the motion of the planets. In 1609, the German astronomer Johannes Kepler announced that the planets travel around the sun in ellipses, rejecting the old idea of circular orbits.

Not so obvious is the point which moves so that its distance from a point (the focus *F*) is the same as its perpendicular distance from a given line (the directrix). In this case we get a parabola. The parabola has a host of useful properties. If a light source is placed at the focus *F*, the emitted light rays are all parallel to *PM*. On the other hand, if TV signals are sent out by a satellite and hit a parabola-shaped receiving dish, they are gathered together at the focus and are fed into the TV set.

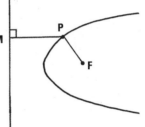

The parabola

If a stick is rotated about a point any fixed point on the stick traces out a circle, but if a point is allowed to moved outwards along the stick in addition to it being rotated this generates a spiral. Pythagoras loved the spiral and much later Leonardo da Vinci spent ten years of his life studying their different types, while René Descartes wrote a treatise on them. The logarithmic spiral is also called the equiangular spiral because it makes the same angle with a radius and the tangent at the point where the radius meets the spiral.

Jacob Bernoulli of the famed mathematical clan from Switzerland was so enamoured with the logarithmic spiral that he wanted it carved on his tomb in Basle. The 'Renaissance man' Emanuel Swedenborg regarded the spiral as the most perfect of shapes. A three-dimensional spiral which winds itself around a cylinder is called a helix. Two of these – a double helix – form the basic structure of DNA.

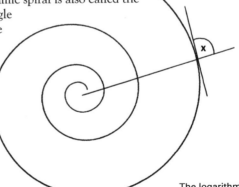

The logarithmic spiral

AD**1704**	**1890**	**1920s**
Newton classifies the cubic curves	Peano proves a solid square is a curve (the space-filling curve)	Menger and Urysohn define curves as part of topology

There are many classical curves, such as the limaçon, the lemniscate and the various ovals. The cardioid derives its name from being shaped like a heart. The catenary curve was the subject of research in the 18th century and it was identified as the curve formed by a chain hanging between two points. The parabola is the curve seen in a suspension bridge hanging between its two vertical pylons.

P

Q

Three-bar motion

One aspect of 19th-century research on curves was on those curves that were generated by mechanical rods. This type of question was an extension of the problem solved approximately by the Scottish engineer James Watt who designed jointed rods to turn circular motion into linear motion. In the steam age this was a significant step forward.

The simplest of these mechanical gadgets is the three-bar motion, where the bars are jointed together with fixed positions at either end. If the 'coupler bar' PQ moves in any which way, the locus of a point on it turns out to be a curve of degree six, a 'sextic curve'.

Algebraic curves Following Descartes, who revolutionized geometry with the introduction of x, y and z coordinates and the Cartesian axes named after him, the conics could now be studied as algebraic equations. For example, the circle of radius 1 has the equation $x^2 + y^2 = 1$, which is an equation of the second degree, as *all* conics are. A new branch of geometry grew up called algebraic geometry.

In a major study Isaac Newton classified curves described by algebraic equations of degree three, or cubic curves. Compared with the four basic conics, 78 types were found, grouped into five classes. The explosion of the number of different types continues for quartic curves, with so many different types that the full classification has never been carried out.

The study of curves as algebraic equations is not the whole story. Many curves such as catenarys, cycloids (curves traced out by a point on a revolving wheel) and spirals are not easily expressible as algebraic equations.

A definition What mathematicians were after was a definition of a curve itself, not just specific examples. Camille Jordan proposed a theory of curves built on the definition of a curve in terms of variable points.

Here's an example. If we let $x = t^2$ and $y = 2t$ then, for different values of t, we get many different points that we can write as coordinates (x, y). For example, if $t = 0$ we get the point $(0, 0)$, $t = 1$ gives the point $(1, 2)$, and so on. If we plot these points on the x–y axes and 'join the dots' we will get a parabola. Jordan refined this idea of points being traced out. For him this was the definition of a curve.

A simple closed Jordan curve

Jordan's curves can be intricate, even when they are like the circle, in that they are 'simple' (do not cross themselves) and 'closed' (have no beginning or end). Jordan's celebrated theorem has meaning. It states that a simple closed curve has an inside and an outside. Its apparent 'obviousness' is a deception.

In Italy, Giuseppe Peano caused a sensation when, in 1890, he showed that, according to Jordan's definition, a filled in square is a curve. He could organize the points on a square so that they could *all* be 'traced out' and at the same time conform to Jordan's definition. This was called a space-filling curve and blew a hole in Jordan's definition – clearly a square is not a curve in the conventional sense.

Examples of space-filling curves and other pathological examples caused mathematicians to go back to the drawing board once more and think about the foundations of curve theory. The whole question of developing a better definition of a curve was raised. At the start of the 20th century this task took mathematics into the new field of topology.

the condensed idea
Going round the bend

23 Topology

Topology is the branch of geometry that deals with the properties of surfaces and general shapes but is unconcerned with the measurement of lengths or angles. High on the agenda are qualities which do not change when shapes are transformed into other shapes. We are allowed to push and pull the shape in any direction and for this reason topology is sometimes described as 'rubber sheet geometry'. Topologists are people who cannot tell the difference between a donut and a coffee cup!

A donut is a surface with a single hole in it. A coffee cup is the same, where the hole takes the form of the handle. Here's how a donut can be transformed into a coffee cup.

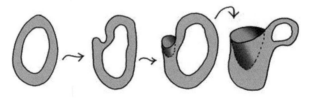

Classifying polyhedra The most basic shapes studied by topologists are polyhedra ('poly' means 'many' and 'hedra' means 'faces'). An example of a polyhedron is a cube, with 6 square faces, 8 vertices (points at the junction of the faces) and 12 edges (the lines joining the vertices). The cube is a *regular* polyhedron because:

- all the faces are the same regular shape,
- all the angles between edges meeting at a vertex are equal.

timeline

Topology is a relatively new subject, but it can still be traced back to the Greeks, and indeed the culminating result of Euclid's *Elements* is to show that there are *exactly* five regular polyhedra. These are the Platonic solids:

Tetrahedron

Cube

- tetrahedron (with 4 triangular faces),
- cube (with 6 square faces),
- octahedron (with 8 triangular faces),
- dodecahedron (with 12 pentagonal faces),
- icosahedron (with 20 triangular faces).

Octahedron

Dodecahedron

If we drop the condition that each face be the same, we are in the realm of the Archimedean solids which are semi-regular. Examples can be generated from the Platonic solids. If we slice off (truncate) some corners of the icosahedron we have the shape used as the design for the modern soccer ball. The 32 faces that form the panels are made up of 12 pentagons and 20 hexagons. There are 90 edges and 60 vertices. It is also the shape of buckminsterfullerene molecules, named after the visionary Richard Buckminster Fuller, creator of the geodesic dome. These 'bucky balls' are a newly discovered form of carbon, C_{60}, with a carbon atom found at each vertex.

Icosahedron

Truncated
icosahedron

Euler's formula Euler's formula is that the number of vertices V, edges E and faces F, of a polyhedron are connected by the formula

$$V - E + F = 2$$

For example, for a cube, $V = 8$, $E = 12$ and $F = 6$ so $V - E + F = 8 - 12 + 6 = 2$ and, for buckminsterfullerene, $V - E + F = 60 - 90 + 32 = 2$. This theorem actually challenges the very notion of a polyhedron.

If a cube has a 'tunnel' through it, is it a real polyhedron? For this shape, $V = 16$, $E = 32$, $F = 16$ and $V - E + F = 16 - 32 + 16 = 0$. Euler's formula does not work. To reclaim the correctness of the formula, the type of

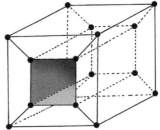

The cube with a tunnel

1858	**1961**	**1982**	**2002**
Möbius and Listing introduce the Möbius strip	Stephen Smale proves the Poincaré conjecture in dimensions greater than 4	Michael Freedman proves the Poincaré conjecture in dimension equal to 4	Perelman proves the Poincaré conjecture for dimension 3

polyhedron could be limited to those *without* tunnels. Alternatively, the formula could be generalized to include this peculiarity.

Classification of surfaces A topologist might regard the donut and the coffee cup as identical but what sort of surface is *different* from the donut? A candidate here is the rubber ball. There is no way of transforming the donut into a ball since the donut has a hole but the ball does not. This is a fundamental difference between the two surfaces. So a way of classifying surfaces is by the number of holes they contain.

Let's take a surface with *r* holes and divide it into regions bounded by edges joining vertices planted on the surface. Once this is done, we can count the number of vertices, edges, and faces. For any division, the Euler expression $V - E + F$ always has the same value, called the Euler characteristic of the surface:

$$V - E + F = 2 - 2r$$

Möbius strip

If the surface has no holes ($r = 0$) as was the case with ordinary polyhedra, the formula reduces to Euler's $V - E + F = 2$. In the case of one hole ($r = 1$), as was the case with the cube with a tunnel, $V - E + F = 0$.

One-sided surfaces Ordinarily a surface will have two sides. The outside of a ball is different from the inside and the only way to cross from one side to the other is to drill a hole in the ball – a cutting operation which is not allowed in topology (you can stretch but you cannot cut). A piece of paper is another example of a surface with two sides. The only place where one side meets the other side is along the bounding curve formed by the edges of the paper.

Klein bottle

The idea of a one-sided surface seems far-fetched. Nevertheless, a famous one was discovered by the German mathematician and astronomer August Möbius in the 19th century. The way to construct such a surface is to take a strip of paper, give it one twist and then stick the ends together. The result is a 'Möbius strip', a one-sided surface with a boundary curve. You can take your pencil and start drawing a line along its middle. Before long you are back where you started!

It is even possible to have a one-sided surface that does not have a boundary curve. This is the 'Klein bottle' named after the German mathematician Felix Klein. What's particularly impressive about this bottle is that it does not intersect itself. However, it is not possible to make a model of the Klein bottle in three-dimensional space without a physical intersection, for it properly lives in four dimensions where it would have no intersections.

Both these surfaces are examples of what topologists call 'manifolds' – geometrical surfaces that look like pieces of two-dimensional paper when *small* portions are viewed by themselves. Since the Klein bottle has no boundary it is called a 'closed' 2-manifold.

The Poincaré conjecture For more than a century, an outstanding problem in topology was the celebrated Poincaré conjecture, named after Henri Poincaré. The conjecture centres on the connection between algebra and topology.

The part of the conjecture that remained unsolved until recently applied to closed 3-manifolds. These can be complicated – imagine a Klein bottle with an extra dimension. Poincaré conjectured that certain closed 3-manifolds which had all the algebraic hallmarks of being three-dimensional spheres actually had to be spheres. It was as if you walked around a giant ball and all the clues you received indicated it was a sphere but because you could not see the big picture you wondered if it really was a sphere.

No one could prove the Poincaré conjecture for 3-manifolds. Was it true or was it false? It had been proven for all other dimensions but the 3-manifold case was obstinate. There were many false proofs, until in 2002 when it was recognized that Grigori Perelman of the Steklov Institute in St Petersburg had finally proved it. Like the solution to other great problems in mathematics, the solution techniques for the Poincaré conjecture lay outside its immediate area, in a technique related to heat diffusion.

the condensed idea
From donuts to coffee cups

24 Dimension

Leonardo da Vinci wrote in his notebook: 'The science of painting begins with the point, then comes the line, the plane comes third, and the fourth the body in its vesture of planes.' In da Vinci's hierarchy, the point has dimension zero, the line is one-dimensional, the plane is two-dimensional and space is three-dimensional. What could be more obvious? It is the way the point, line, plane and solid geometry had been propagated by the Greek geometer Euclid, and Leonardo was following Euclid's presentation.

That physical space is three-dimensional has been the view for millennia. In physical space we can move *out* of this page along the x-axis, or *across* it horizontally along the y-axis or vertically *up* the z-axis, or any combination of these. Relative to the origin (where the three axes meet) every point has a set of spatial coordinates specified by values of x, y and z and written in the form (x, y, z).

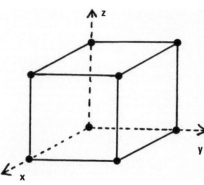

A cube plainly has these three dimensions and so does everything else which has solidity. At school we are normally taught the geometry of the plane which is two-dimensional and we then move up to three dimensions – to 'solid geometry' – and stop there.

Around the beginning of the 19th century, mathematicians began to dabble in four dimensions and in even higher n-dimensional mathematics. Many philosophers and mathematicians began to ask whether higher dimensions existed.

The space of three dimensions

Higher physical dimensions Many leading mathematicians in the past thought that four dimensions could not be imagined. They queried the reality of four dimensions, and it became a challenge to explain this.

timeline

c.300BC	AD1877
Euclid describes a three-dimensional world	Cantor is surprised by his controversial discoveries in dimension theory

A common way to explain why four dimensions could be possible was to fall back to two dimensions. In 1884, an English schoolmaster and theologian, Edwin Abbott, published a highly popular book about 'flatlanders' who lived in the two-dimensional plane. They could not see triangles, squares or circles which existed in Flatland because they could not go out into the third dimension to view them. Their vision was severely limited. They had the same problems thinking about a third dimension that we do thinking of a fourth. But reading Abbott puts us into the frame of mind to accept the fourth dimension.

The need to contemplate the actual existence of a four-dimensional space became more urgent when Einstein came on the scene. Four-dimensional geometry became more plausible, and even understandable, because the extra dimension in Einstein's model is *time*. Unlike Newton, Einstein conceived time as bound together with space in a four-dimensional space–time continuum. Einstein decreed that we live in a four-dimensional world with four coordinates (x, y, z, t) where t designates time.

Nowadays the four-dimensional Einsteinian world seems quite tame and matter of fact. A more recent model of physical reality is based on 'strings'. In this theory, the familiar subatomic particles like electrons are the manifestations of extremely tiny vibrating strings. String theory suggests a replacement of the four-dimensional space–time continuum by a higher-dimensional version. Current research suggests that the dimension of the accommodating space–time continuum for string theory should be either 10, 11 or 26, depending on further assumptions and differing points of view.

A huge 2000 tonne magnet at CERN near Geneva, Switzerland, designed to engineer collisions of particles at high speeds, might help to resolve the issue. It is intended to uncover the structure of matter and, as a by-product, may point to a better theory and the 'correct' answer on dimensionality. The smart money seems to be that we're living in an 11-dimensional universe.

Hyperspace Unlike higher physical dimensions, there is absolutely no problem with a *mathematical space* of more than three dimensions. Mathematical space can be any number of dimensions. Since the early 19th century mathematicians have habitually used n variables in their work. George Green, a miller from Nottingham who explored the mathematics of electricity, and pure mathematicians A.L. Cauchy, Arthur Cayley and Hermann Grassmann, all described their mathematics in terms of n-dimensional hyperspace. There seemed no good reason to limit the mathematics and everything to be gained in elegance and clarity.

The idea behind n dimensions is merely an extension of three-dimensional coordinates (x, y, z) to an unspecified number of variables. A circle in two dimensions has an equation $x^2 + y^2 = 1$, a sphere in three dimensions has an equation $x^2 + y^2 + z^2 = 1$, so why not a hypersphere in four dimensions with equation $x^2 + y^2 + z^2 + w^2 = 1$.

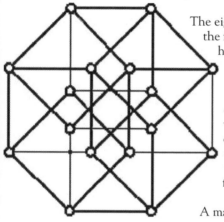

The four-dimensional cube

The eight corners of a cube in three dimensions have coordinates of the form (x, y, z) where each of the x, y, z are either 0 or 1. The cube has six faces each of which is a square and there are $2 \times 2 \times 2 = 8$ corners. What about a four-dimensional cube? It will have coordinates of the form (x, y, z, w) where each of the x, y, z and w are either 0 or 1. So there are $2 \times 2 \times 2 \times 2 = 16$ possible corners for the four-dimensional cube, and eight faces, each of which is a cube. We cannot actually see this four-dimensional cube but we can create an artist's impression of it on this sheet of paper. This shows a projection of the four-dimensional cube which exists in the mathematician's imagination. The cubic faces can just about be perceived.

A mathematical space of many dimensions is quite a common occurrence for pure mathematicians. No claim is made for its actual existence though it may be assumed to exist in an ideal Platonic world. In the great problem of the classification of groups, for instance (see page 155), the 'monster group' is a way of measuring symmetry in a mathematical space of 196,883 dimensions. We cannot 'see' this space in the same way as we can in the ordinary three-dimensional space, but it can still be imagined and dealt with in a precise way by modern algebra.

The mathematician's concern for dimension is entirely separate from the meaning the physicist attaches to dimensional analysis. The common units of physics are measured in terms of mass M, length L, and time T. So, using their dimensional analysis a physicist can check whether equations make sense since both sides of an equation must have the same dimensions.

It is no good having force = velocity. A dimensional analysis gives velocity as metres per second so it has dimension of length divided by time or L/T, which we write as LT^{-1}. Force is mass times acceleration, and as acceleration is metres per second per second, the net result is that force will have dimensions MLT^{-2}.

Topology Dimension theory is part of general topology. Other concepts of dimension can be defined independently in terms of abstract mathematical spaces. A major task is to show how they relate to each other. Leading figures in many branches of mathematics have delved into the meaning of dimension including Henri Lebesgue, L.E.J. Brouwer, Karl Menger, Paul Urysohn and Leopold Vietoris (until recently the oldest person in Austria, who died in 2002 aged 110).

The pivotal book on the subject was *Dimension Theory*. Published in 1948 by Witold Hurewicz and Henry Wallman – it is still seen as a watershed in our understanding of the concept of dimension.

Coordinated people

Human beings themselves are many dimensioned things. A human being has many more 'coordinates' than three. We could use (*a, b, c, d, e, f, g, h*), for age, height, weight, gender, shoe size, eye colour, hair colour, nationality, and so on. In place of geometrical points we might have people. If we limit ourselves to this eight-dimensional 'space' of people, John Doe might have coordinates like (43 years, 165 cm, 83 kg, male, 9, blue, blond, Danish) and Mary Smith's coordinates might be (26 years, 157 cm, 56 kg, female, 4, brown, brunette, British).

Dimension in all its forms From the three dimensions introduced by the Greeks the concept of dimension has been critically analysed and extended.

The n dimensions of mathematical space were introduced quite painlessly, while physicists have based theories on space–time (of dimension four) and recent versions of string theory (see page 97) which demand, 10, 11 and 26 dimensions. There have been forays into fractional dimensions with fractal shapes (see page 100) with several different measures being studied. Hilbert introduced an infinite-dimensional mathematical space that is now a basic framework for pure mathematicians. Dimension is so much more than the one, two, three of Euclidean geometry.

the condensed idea
Beyond the third dimension

25 Fractals

In March 1980, the state-of-the-art mainframe computer at the IBM research centre at Yorktown Heights, New York State, was issuing its instructions to an ancient Tektronix printing device. It dutifully struck dots in curious places on a white page, and when it had stopped its clatter the result looked like a handful of dust smudged across the sheet. Benoît Mandelbrot rubbed his eyes in disbelief. He saw it was important, but what was it? The image that slowly appeared before him was like the black and white print emerging from a photographic developing bath. It was a first glimpse of that icon in the world of fractals – the Mandelbrot set.

This was experimental mathematics par excellence, an approach to the subject in which mathematicians had their laboratory benches just like the physicists and chemists. They too could now do experiments. New vistas opened up – literally. It was a liberation from the arid climes of 'definition, theorem, proof', though a return to the rigours of rational argument would have to come albeit later.

The downside of this experimental approach was that the visual images preceded a theoretical underpinning. Experimentalists were navigating without a map. Although Mandelbrot coined the word 'fractals', what were they? Could there be a precise definition for them in the usual way of mathematics? In the beginning, Mandelbrot didn't want to do this. He didn't want to destroy the magic of the experience by honing a sharp definition which might be inadequate and limiting. He felt the notion of a fractal, 'like a good wine – demanded a bit of aging before being "bottled".'

The Mandelbrot set Mandelbrot and his colleagues were not being particularly abstruse mathematicians. They were playing with the simplest of formulae. The whole idea is based on iteration – the practice of applying a

timeline

AD1879

Cayley works on a precursor of
modern fractals

1904

von Koch creates hi
snowflake curve

formula time and time again. The formula which generated the Mandelbrot set was simply $x^2 + c$.

The first thing we do is choose a value of c. Let's choose $c = 0.5$. Starting with $x = 0$ we substitute into the formula $x^2 + 0.5$. This first calculation gives 0.5 again. We now use this as x, substituting it into $x^2 + 0.5$ to give a second calculation: $(0.5)^2 + 0.5 = 0.75$. We keep going, and at the third stage this will be $(0.75)^2 + 0.5 = 1.0625$. All these calculations can be done on a handheld calculator. Carrying on we find that the answer gets bigger and bigger.

Let's try another value of c, this time $c = -0.5$. As before we start at $x = 0$ and substitute it into $x^2 - 0.5$ to give -0.5. Carrying on we get -0.25, but this time the values do not become bigger and bigger but, after some oscillations, settle down to a figure near $-0.3660...$

So by choosing $c = 0.5$ the sequence starting at $x = 0$ zooms off to infinity, but by choosing $c = -0.5$ we find that the sequence starting at $x = 0$ actually converges to a value near -0.3660. The Mandelbrot set consists of all those values of c for which the sequence starting at $x = 0$ does *not* escape to infinity.

The Mandelbrot set

This is not the whole story because so far we have only considered the one-dimensional real numbers – giving a one-dimensional Mandelbrot set so we wouldn't see much. What needs to be considered is the same formula $z^2 + c$ but with z and c as two-dimensional complex numbers (see page 32). This will give us a two-dimensional Mandelbrot set.

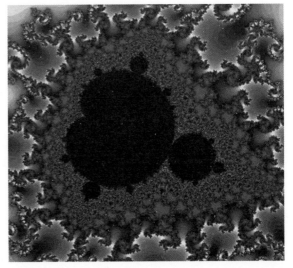

For some values of c in the Mandelbrot set, the sequence of zs may do all sorts of strange things like dance between a number of points but they will not escape to infinity. In the Mandelbrot set we see another key property of fractals, that of self-similarity. If you zoom into the set you will not be sure of the level of magnification because you will just see more Mandelbrot sets.

Before Mandelbrot Like most things in mathematics, discoveries are rarely brand new. Looking into the history Mandelbrot found that mathematicians such as Henri Poincaré and Arthur Cayley had brief glimmerings of the idea a hundred years before him. Unfortunately they did not have the computing power to investigate matters further.

The shapes discovered by the first wave of fractal theorists included crinkly curves and the 'monster curves' that had previously been dismissed as pathological examples of curves. As they were so pathological they had been locked up in the mathematician's cupboard and given little attention. What was wanted then were the more normal 'smooth' curves which could be dealt with by the differential calculus. With the popularity of fractals, other mathematicians whose work was resurrected were Gaston Julia and Pierre Fatou who worked on fractal-like structures in the complex plane in the years following the First World War. Their curves were not called fractals, of course, and they did not have the technological equipment to see their shapes.

The generating element of the Koch snowflake

Other famous fractals The famous Koch curve is named after the Swedish mathematician Niels Fabian Helge von Koch. The snowflake curve is practically the first fractal curve. It is generated from the side of the triangle treated as an element, splitting it into three parts each of length ⅓ and adding a triangle in the middle position.

The curious property of the Koch curve is that it has a finite area, because it always stays within a circle, but at each stage of its generation its length increases. It is a curve which encloses a finite area but has an 'infinite' circumference!

Another famous fractal is named after the Polish mathematician Wacław Sierpiński. It is found by subtracting triangles from an equilateral triangle; and by continuing this process, we get the Sierpiński gasket (generated by a different process on page 54).

The Koch snowflake

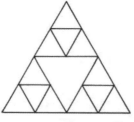

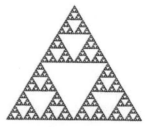

The Sierpiński gasket

Fractional dimension The way Felix Hausdorff looked at dimension was innovative. It has to do with scale. If a line is scaled up by a factor of 3 it is 3 times longer than it was previously. Since this $3 = 3^1$ a line is said to have dimension 1. If a solid square is scaled up by a factor of 3 its *area* is 9 times its previous value or 3^2 and so the dimension is 2. If a cube is scaled up by this factor its volume is 27 or 3^3 times its previous value, so its dimension is 3. These values of the Hausdorff dimension all coincide with our expectations for a line, square, or cube.

If the basic unit of the Koch curve is scaled up by 3, it becomes 4 times longer than it was before. Following the scheme described, the Hausdorff dimension is the value of D for which $4 = 3^D$. An alternative calculation is that

$$D = \frac{\log 4}{\log 3}$$

which means that D for the Koch curve is approximately 1.262. With fractals it is frequently the case that the Hausdorff dimension is greater than the ordinary dimension, which is 1 in the case of Koch curve.

The Hausdorff dimension informed Mandelbrot's definition of a fractal – a set of points whose value of D is not a whole number. Fractional dimension became the key property of fractals.

The applications of fractals The potential for the applications of fractals is wide. Fractals could well be the mathematical medium which models such natural objects as plant growth, or cloud formation.

Fractals have already been applied to the growth of marine organisms such as corals and sponges. The spread of modern cities has been shown to have a similarity with fractal growth. In medicine they have found application in the modelling of brain activity. And the fractal nature of movements of stocks and shares and the foreign exchange markets has also been investigated. Mandelbrot's work opened up a new vista and there is much still to be discovered.

the condensed idea
Shapes with fractional dimension

26 Chaos

How is it possible to have a theory of chaos? Surely chaos happens in the absence of theory? The story goes back to 1812. While Napoleon was advancing on Moscow, his compatriot the Marquis Pierre-Simon de Laplace published an essay on the deterministic universe: if at one particular instant, the positions and velocities of all objects in the universe were known, and the forces acting on them, then these quantities could be calculated exactly for all future times. The universe and all objects in it would be completely determined. Chaos theory shows us that the world is more intricate that that.

In the real world we cannot know all the positions, velocities and forces *exactly*, but the corollary to Laplace's belief was that if we knew approximate values at one instant, the universe would not be much different anyway. This was reasonable, for surely sprinters who started a tenth of a second after the gun had fired would break the tape only a tenth of a second off their usual time. The belief was that small discrepancies in initial conditions meant small discrepancies in outcomes. Chaos theory exploded this idea.

The butterfly effect The butterfly effect shows how initial conditions slightly different from the given ones, can produce an actual result very different from the predictions. If fine weather is predicted for a day in Europe, but a butterfly flaps its wings in South America then this could actually presage storms on the other side of the world – because the flapping of the wings changes the air pressure very slightly causing a weather pattern completely different from the one originally forecast.

We can illustrate the idea with a simple mechanical experiment. If you drop a ball-bearing through the opening in the top of a pinboard box it will progress downwards, being deflected one way or the other by the different pins it

Laplace publishes his essay on a deterministic world

Poncaré encounters chaos in his work on the three-body problem for which he is awarded a prize by King Oscar of Sweden

encounters on route until it reaches a finishing slot at the bottom. You might then attempt to let another identical ball-bearing go from the very same position with exactly the same velocity. If you could do this *exactly* then the Marquis de Laplace would be correct and the path followed by the ball would be exactly the same. If the first ball dropped into the third slot from the right, then so would the second ball.

But of course you cannot let the ball go from exactly the same position with exactly the same velocity and force. In reality, there will be a very slight difference which you might not even be able to measure. The result is the ball-bearing may take a very different route to the bottom and probably end up in a different slot.

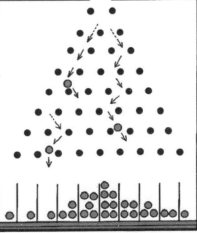

Pinboard box experiment

A simple pendulum The free pendulum is one of the simplest mechanical systems to analyse. As the pendulum swings back and forth, it gradually loses energy. The displacement from the vertical and the (angular) velocity of the bob decrease until it is eventually stationary.

The movement of the bob can be plotted in a phase diagram. On the horizontal axis the (angular) displacement is measured and on the vertical axis the velocity is measured. The point of release is plotted at the point A on the positive horizontal axis. At A the displacement is at a maximum and the velocity is zero. As the bob moves through the vertical axis (where the displacement is zero) the velocity is at a maximum, and this is plotted on the phase diagram at B. At C when the bob is at the other extremity of its swing, the displacement is negative and the velocity is zero. The bob then swings back through D (where it is moving in the opposite direction so its velocity is negative) and completes one swing at E. In the phase diagram this is represented by a rotation through 360 degrees, but because the swing is reduced the point E is shown *inside* A. As the pendulum swings less and less this phase portrait spirals into the origin until eventually the pendulum comes to rest.

The free pendulum

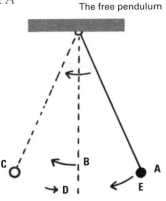

961

1971

2004

orenz observes the
utterfly effect

Robert May investigates chaos
in the population model

Chaos theory reaches popular culture
in the film *The Butterfly Effect*

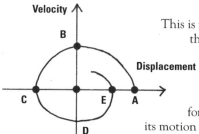

Phase diagram for the simple pendulum

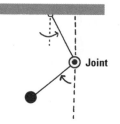

Joint

Movement of double pendulum

This is not the case for the double pendulum in which the bob is at the end of a jointed pair of rods. If the displacement is small the motion of the double pendulum is similar to the simple pendulum, but if the displacement is large the bob swings, rotates, and lurches about and the displacement about the intermediate joint is seemingly random. If the motion is not forced, the bob will also come to rest but the curve that describes its motion is far from the well-behaved spiral of the single pendulum.

Chaotic motion The characteristic of chaos is that a deterministic system may appear to generate random behaviour. Let's look at another example, the repeating, or iterative, formula $a \times p \times (1 - p)$ where p stands for the population, measured as a proportion on a scale from 0 to 1. The value of a must be somewhere between 0 and 4 to guarantee that the value of p stays in the range from 0 to 1.

Let's model the population when $a = 2$. If we pick a starting value of, say $p = 0.3$ at *time* = 0, then to find the population at *time* = 1, we feed $p = 0.3$ into $a \times p \times (1 - p)$ to give 0.42. Using only a handheld calculator we can repeat this operation, this time with $p = 0.42$, to give us the next figure (0.4872). Progressing in this way, we find the population at later times. In this case, the population quickly settles down to $p = 0.5$. This settling down always takes place for values of a less than 3.

If we now choose $a = 3.9$, a value near the maximum permissible, and use the same initial population $p = 0.3$ the population does *not* settle down but oscillates wildly. This is because the value of a is in the 'chaotic region', that is, a is a number greater than 3.57. Moreover, if we choose a different initial population, $p = 0.29$, a value close to 0.3, the population growth shadows the previous growth pattern for the first few steps but then starts to diverge from it completely. This is the behaviour experienced by Edward Lorenz in 1961 (see box).

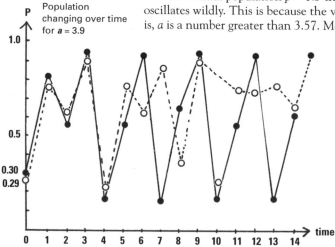

Population changing over time for *a* = 3.9

Forecasting the weather
Even with very powerful computers we all know that we cannot forecast the weather more than a few days in advance. Over just a few days forecasting the weather still gives us

nasty surprises. This is because the equations which govern the weather are nonlinear – they involve the variables multiplied together, not just the variables themselves.

The theory behind the mathematics of weather forecasting was worked out independently by the French engineer Claude Navier in 1821 and the British mathematical physicist George Gabriel Stokes in 1845. The Navier–Stokes equations that resulted are of intense interest to scientists. The Clay Mathematics Institute in Cambridge, Massachusetts has offered a million dollar prize to whoever makes substantial progress towards a mathematical theory that unlocks their secrets. Applied to the problem of fluid flow, much is known about the steady movements of the upper atmosphere. But air flow near the surface of the Earth creates turbulence and chaos results, with the subsequent behaviour largely unknown.

While a lot is known about the theory of linear systems of equations, the Navier–Stokes equations contain nonlinear terms which make them intractable. Practically the only way of solving them is to do so numerically by using powerful computers.

From meteorology to mathematics

The discovery of the butterfly effect happened by chance around 1961. When meteorologist Edward Lorenz at MIT went to have a cup of coffee and left his ancient computer plotting away he came back to something unexpected. He had been aiming to recapture some interesting weather plots but found the new graph unrecognizable. This was strange for he had entered in the same initial values and the same picture should have been drawn out. Was it time to trade in his old computer and get something more reliable?

After some thought he did spot a difference in the way he had entered the initial values: before he had used six decimal places but on the rerun he only bothered with three. To explain the disparity he coined the term 'butterfly effect'. After this discovery his intellectual interests migrated to mathematics.

Strange attractors Dynamic systems can be thought of possessing 'attractors' in their phase diagrams. In the case of the simple pendulum the attractor is the single point at the origin that the motion is directed towards. With the double pendulum it's more complicated, but even here the phase portrait will display some regularity and be attracted to a set of points in the phase diagram. For systems like this the set of points may form a fractal (see page 100) which is called a 'strange' attractor that will have a definite mathematical structure. So all is not lost. In the new chaos theory, it is not so much 'chaotic' chaos that results as 'regular' chaos.

the condensed idea
The wildness of regularity

27 The parallel postulate

This dramatic story begins with a simple geometric scenario. Imagine a line *l* and a point *P* not on the line. How many lines can we draw through *P* parallel to the line *l*? It appears obvious that there is exactly one line through *P* which will never meet *l* no matter how far it is extended in either

direction. This seems self-evident and in perfect agreement with common sense. Euclid of Alexandria included a variant of it as one of his postulates in that foundation of geometry, the *Elements*.

Common sense is not always a reliable guide. We shall see whether Euclid's assumption makes mathematical sense.

Euclid's *Elements* Euclid's geometry is set out in the 13 books of the *Elements*, written around 300 BC. One of the most influential mathematics texts ever written, Greek mathematicians constantly referred to it as the first systematic codification of geometry. Later scholars studied and translated it from extant manuscripts and it was handed down and universally praised as the very model of what geometry should be.

The *Elements* percolated down to school level and readings from the 'sacred book' became the way geometry was taught. It proved unsuitable for the youngest pupils, however. As the poet A.C. Hilton quipped: 'though they wrote it all by rote, they did not write it right'. You might say Euclid was written for men not boys. In English schools, it reached the zenith of its

timeline

c.300BC

Euclid includes the parallel postulate in his *Elements*

1829–31

Lobachevsky and Bolyai publish their work on hyperbolic geometry

influence as a subject in the curriculum during the 19th century but it remains a touchstone for mathematicians today.

It is the style of Euclid's *Elements* that makes it noteworthy – its achievement is the presentation of geometry as a sequence of proven propositions. Sherlock Holmes would have admired its deductive system which advanced logically from the clearly stated postulates and may have castigated Dr Watson for not seeing it as a 'cold unemotional system'.

While the edifice of Euclid's geometry rests on the postulates (what are now called axioms; see box), these were not enough. Euclid added 'definitions' and 'common notions'. The definitions include such declarations as 'a *point* is that which has no part' and 'a *line* is breadthless length'. Common notions include such items as 'the whole is greater than the part' and 'things which are equal to the same thing are also equal to one another'. It was only towards the end of the 19th century that it was recognized that Euclid had made tacit assumptions.

Euclid's postulates

One of the characteristics of mathematics is that a few assumptions can generate extensive theories. Euclid's postulates are an excellent example, and one that set the model for later axiomatic systems. His five postulates are:

1. A straight line can be drawn from any point to any point.
2. A finite straight line can be extended continuously in a straight line.
3. A circle can be constructed with any centre and any radius.
4. All right angles are equal to each other.
5. If a straight line falling on two straight lines makes the interior angles on the same side less than two right angles, the two straight lines, if extended indefinitely, meet on that side on which the angles are less than two right angles.

The fifth postulate It is Euclid's fifth postulate that caused controversy over 2000 years after the *Elements* first appeared. In style alone, it looks out of place through its wordiness and clumsiness. Euclid himself was unhappy with it but he needed it to prove propositions and had to include it. He tried to prove it from the other postulates but failed.

Later mathematicians either tried to prove it or replace it by a simpler postulate. In 1795, John Playfair stated it in a form which gained popularity: for a line *l* and a point *P* not on the line *l* there is a *unique* line passing through *P* parallel to *l*. Around the same time, Adrien Marie Legendre substituted another equivalent version when he asserted the existence of

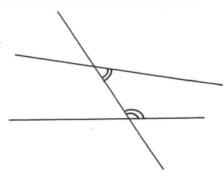

a triangle whose angles add up to 180 degrees. These new forms of the fifth postulate went some way to meet the objection of artificiality. They were more acceptable than the cumbersome version given by Euclid.

Another line of attack was to search for the elusive proof of the fifth postulate. This exerted a powerful attraction on its adherents. If a proof could be found, the postulate would become a theorem and it could retire from the firing line. Unfortunately attempts to do this turned out to be excellent examples of circular reasoning, arguments which assume the very thing they are trying to prove.

Non-Euclidean geometry A breakthrough came through the work of Carl Friedrich Gauss, János Bolyai and Nikolai Ivanovich Lobachevsky. Gauss did not publish his work, but it seems clear he reached his conclusions in 1817. Bolyai published in 1831 and Lobachevsky, independently, in 1829, causing a prority dispute between these two. There is no doubting the brilliance of all these men. They effectively showed that the fifth postulate was independent of the other four postulates. By adding its negation to the other four postulates, they showed a consistent system was possible.

Bolyai and Lobachevsky constructed a new geometry by allowing there to be *more* than one line through P that does not meet the line *l*. How can this be? Surely the dotted lines meet *l*. If we accept this we are unconsciously falling in with Euclid's view. The diagram is therefore a confidence trick, for what Bolyai and Lobachevsky were proposing was a new sort of geometry which does not conform to the commonsense one of Euclid. In fact, their non-Euclidean geometry can be thought of as the geometry on the curved surface of what is known as a pseudosphere.

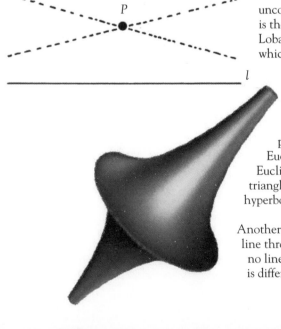

The shortest paths between the points on a pseudosphere play the same role as straight lines in Euclid's geometry. One of the curiosities of this non-Euclidean geometry is that the sum of the angles in a triangle is *less* than 180 degrees. This geometry is called hyperbolic geometry.

Another alternative to the fifth postulate states that *every* line through P meets the line *l*. Put a different way, there are no lines through P which are 'parallel' to *l*. This geometry is different from Bolyai's and Lobachevsky's, but it is a genuine geometry nevertheless. One model for it is the geometry on the surface of a sphere. Here the great circles (those circles that have the same circumference as the sphere itself) play the role of straight lines in Euclidean geometry. In this non-Euclidean geometry the sum of the angles in a triangle is *greater* than 180 degrees. This is called elliptic geometry and is associated with the German mathematician Benhard Riemann who investigated it in the 1850s.

The geometry of Euclid which had been thought to be the one true geometry – according to Immanuel Kant, the geometry 'hard-wired into man' – had been knocked off its pedestal. Euclidean geometry was now one of many systems, sandwiched between hyperbolic and elliptic geometry. The different versions were unified under one umbrella by Felix Klein in 1872. The advent of non-Euclidean geometry was an earth-shaking event in mathematics and paved the way to the geometry of Einstein's general relativity (see page 192). It is the *general* theory of relativity which demands a new kind of geometry – the geometry of curved space–time, or Riemannian geometry. It was this non-Euclidean geometry which now explained why things fall down, and not Newton's attractive gravitational force between objects. The presence of massive objects in space, like the Earth and the Sun cause space–time to be curved. A marble on a sheet of thin rubber will cause a small indentation, but try placing a bowling ball on it and a great warp will result.

This curvature measured by Riemannian geometry predicts how light beams bend in the presence of massive space objects. Ordinary Euclidean space, with time as an independent component, will not suffice for general relativity. One reason is that Euclidean space is flat – there is no curvature. Think of a sheet of paper lying on a table; we can say that at any point on the paper the curvature is zero. Underlying Riemannian space–time is a concept of curvature which varies continuously – just as the curvature of a rumpled piece of cloth varies from point to point. It's like looking in a bendy fairground mirror – the image you see depends on where you look in the mirror.

No wonder that Gauss was so impressed by young Riemann in the 1850s and even suggested then that the 'metaphysics' of space would be revolutionized by his insights.

the condensed idea
What if parallel lines do meet?

28 Discrete geometry

Geometry is one of the oldest crafts – it literally means earth [*geo*] measuring [*metry*]. In ordinary geometry there are continuous lines and solid shapes to investigate, both of which can be thought of as being composed of points 'next to' each other. Discrete mathematics deals with whole numbers as opposed to the continuous real numbers. Discrete geometry can involve a finite number of points and lines or lattices of points – the continuous is replaced by the isolated.

A lattice or grid is typically the set of points whose coordinates are whole numbers. This geometry poses interesting problems and has applications in such disparate areas as coding theory and the design of scientific experiments.

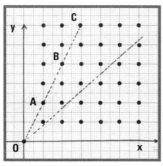

The lattice points of the x/y axes

Let's look at a lighthouse throwing out a beam of light. Imagine the light ray starts at the origin O and sweeps between the horizontal and the vertical. We can ask which rays hit which lattice points (which might be boats tied up in the harbour in a rather uniform arrangement).

The equation of the ray through the origin is $y = mx$. This is the equation of a straight line passing through the origin with gradient m. If the ray is $y = 2x$ then it will hit the point with coordinates $x = 1$ and $y = 2$ because these values satisfy the equation. If the ray hits a lattice point with $x = a$ and $y = b$ the gradient m is the fraction b/a. Consequently if m is not a genuine fraction (it may be $\sqrt{2}$, for example) the light ray will miss all the lattice points.

The light ray $y = 2x$ hits the point A with coordinates $x = 1$ and $y = 2$ but it will not strike the point B with coordinates $x = 2$ and $y = 4$ and all other

timeline

AD 1639

Pascal discovers his theorem while only 16 years old

1806

Brianchon discovers the dual theorem of Pascal's theorem

points 'behind' A (such as C, with coordinates $x = 3$ and $y = 6$, and D with $x = 4$ and $y = 8$) will be obscured. We could imagine ourselves at the origin identifying the points that can be seen from there, and those that are obscured.

We can show that those points with coordinates with $x = a$ and $x = b$ which can be seen are those that are relatively prime to each other. These are points with coordinates, such as $x = 2$ and $y = 3$, where no number other than 1 divides both x and y. The points behind this one will be multiples, such as $x = 4$ and $y = 6$, or $x = 6$ and $y = 9$, and so on.

The points ○ 'visible' from the origin, and the obscured points ×

Pick's theorem

The Austrian mathematician Georg Pick has two claims to fame. One is that he was a close friend of Albert Einstein and proved instrumental in bringing the young scientist to the German University in Prague in 1911. The other is that he wrote a short paper, published in 1899, on 'reticular' geometry. From a lifelong work covering a wide range of topics he is remembered for the captivating Pick's theorem – and what a theorem it is!

Pick's theorem gives a means for computing the area enclosed by a many-sided (or polygonal) shape formed by joining up points whose coordinates are whole numbers. This is pinball mathematics.

To find the area of the shape we shall have to count the number of points ● on the boundary and the number of interior points ○. In our example, the number of points on the boundary is $b = 22$ and the number of interior points is $c = 7$. This is all we need to use Pick's theorem:

$$\text{area} = \tfrac{b}{2} + c - 1$$

From this formula, the area is $\tfrac{22}{2} + 7 - 1 = 17$. The area is 17 square units. It is as simple as that. Pick's theorem can be applied to any shape which joins discrete points with whole number coordinates, the only condition being that the boundary does not cross itself.

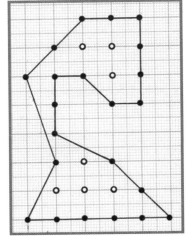

A many-sided or polygonal shape

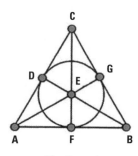

The Fano plane

The Fano plane

The Fano plane geometry was discovered at about the same time as Pick's formula, but has nothing to do with measuring anything at all. Named after the Italian mathematician Gino Fano, who pioneered the study of finite geometry, the Fano plane is the simplest example of a 'projective' geometry. It has only seven points and seven lines.

The seven points are labelled A, B, C, D, E, F and G. It is easy to pick out six of the seven lines but where is the seventh? The properties of the geometry and the way the diagram is constructed make it necessary to treat the seventh line as DFG – the circle passing through D, F and G. This is no problem since lines in discrete geometry do not have to be 'straight' in the conventional sense.

This little geometry has many properties, for example:

- every pair of points determines one line passing through both,
- every pair of lines determines one point lying on both.

These two properties illustrate the remarkable duality which occurs in geometries of this kind. The second property is just the first with the words 'point' and 'line' swapped over, and likewise the first is just the second with the same swaps.

If, in any true statement, we swap the two words and make small adjustments to correct the language, we get another true statement. Projective geometry is very symmetrical. Not so Euclidean geometry. In Euclidean geometry there are parallel lines, that is pairs of lines which never meet. We can quite happily speak of the concept of parallelism in Euclidean geometry. This is not possible in projective geometry. In projective geometry all pairs of lines meet in a point. For mathematicians this means Euclidean geometry is an inferior sort of geometry.

If we remove one line *and* its points from the Fano plane we are once more back in the realm of unsymmetrical Euclidean geometry and the existence of parallel lines. Suppose we remove the 'circular' line DFG to give a Euclidean diagram.

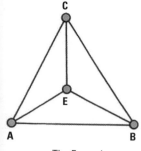

The Fano plane
made Euclidean

With one line fewer there are now six lines: AB, AC, AE, BC, BE and CE. There are now pairs of lines which are 'parallel', namely AB and CE, AC and BE, and BC and AE. Lines are parallel in this sense if they have no points in common – like the lines AB and CE.

The Fano plane occupies an iconic position in mathematics because of its connection to so many ideas and applications. It is one key to Thomas

A	F	B
B	G	C
C	A	D
D	B	E
E	C	F
F	D	G
G	E	A

Kirkman's schoolgirl problem (see page 167). In the theory of designing experiments the Fano plane appears as a protean example, a Steiner Triple System (STS). Given a finite number of n objects an STS is a way of dividing them into blocks of three so that every pair taken from the n objects is in exactly one block. Given the seven objects A, B, C, D, E, F and G the blocks in the STS correspond to the lines of the Fano plane.

A pair of theorems Pascal's theorem and Brianchon's theorem lie on the boundary between continuous and discrete geometry. They are different but related to each other. Pascal's theorem was discovered by Blaise Pascal in 1639 when he was only 16 years old. Let's take a stretched out circle called an ellipse (see page 89) and mark six points along it that we'll call A_1, B_1 and C_1 and A_2, B_2 and C_2. We'll call P the point where the line A_1B_2 intersects A_2B_1; Q the point where the line A_1C_2 intersects A_2C_1; and R the point where the line B_1C_2 intersects B_2C_1. The theorem states that the points P, Q and R all lie on a single straight line.

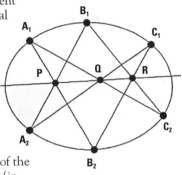

Pascal's theorem

Pascal's theorem is true whatever the positions of the different points around the ellipse. In fact, we could substitute a different conic in place of the ellipse, such as hyperbola, circle, parabola or even a pair of straight lines (in which case the configuration is referred to as 'cat's cradle') and the theorem would still be true.

Brianchon's theorem was discovered much later by the French mathematician and chemist Charles-Julien Brianchon. Let's draw six tangents that we'll call the lines a_1, b_1 and c_1 and a_2, b_2 and c_2, touching the circumference of an ellipse. Next we can define three diagonals, the lines p, q and r, by the meeting of lines, so that: p is the line between the points where a_1 meets b_2 and where a_2 meets b_1; q is the line between the points where a_1 meets c_2 and a_2 meets c_1; and r is the line between the points where b_1 meets c_2 and b_2 meets c_1. Brianchons's theorem states that lines p, q and r meet at a point.

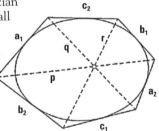

Brianchon's theorem

These two theorems are dual to each other, and it is another instance of the theorems of projective geometry occurring in pairs.

the condensed idea
Individual points of interest

29 Graphs

There are two types of graphs in mathematics. At school we draw curves which show the relationship between variables *x* and *y*. In the other more recent sort, dots are joined up by wiggly lines.

Königsberg is a city in East Prussia famous for the seven bridges which cross the River Pregel. Home to the illustrious philosopher Immanuel Kant, the city and its bridges are also linked with the famous mathematician Leonhard Euler.

In the 18th century a curious question was posed: was it possible to set off and walk around Königsberg crossing each bridge exactly once? The walk does not require us to finish where we started – only that we cross each bridge once.

In 1735, Euler presented his solution to the Russian Academy, a solution which is now seen as the beginning of modern graph theory. In our semi-abstract diagram, the island in the middle of the river is labelled *I* and the banks of the river by *A*, *B* and C. Can you plan a walk for a Sunday afternoon that crosses each bridge just once? Pick up a pencil and try it. The key step is to peel away the semi-abstractness and progress to complete abstraction. In so doing a graph of points and lines is obtained. The land is represented by 'points' and the bridges joining them are represented by 'lines'. We don't care that the lines are not straight or that they differ in length. These things are unimportant. It is only the connections that matter.

Euler made an observation about a successful walk. Apart from the beginning and the end of the walk, every time a bridge is crossed onto a piece of land it must be possible to leave it on a bridge not previously walked over.

timeline

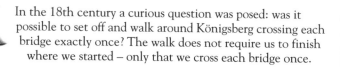

Translating this thought into the abstract picture, we may say that lines meeting at a point must occur in pairs. Apart from two points representing the start and finish of the walk, the bridges can be traversed if and only if each point has an even number of lines incident on it.

The number of lines meeting at a point is called the 'degree' of the point.

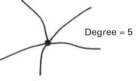

Degree = 5

Euler's theorem states that

> *The bridges of a town or city may be traversed exactly once if, apart from at most two, all points have even degree.*

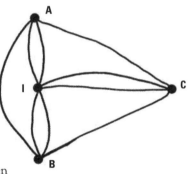

Looking at the graph representing Königsberg, every point is of odd degree. This means that a walk crossing each bridge only once is not possible in Königsberg. If the bridge setup were changed then such a walk may become possible. If an extra bridge were built between the island *I* and *C* the degrees at *I* and *C* would both be even. This means we could begin a walk on *A* and end on *B* having walked over every bridge exactly once. If yet another bridge were built, this time between *A* and *B* (shown right), we could start anywhere and finish at the *same* place because *every* point would have even degree in this case.

The hand-shaking theorem If we were asked to draw a graph that contained three points of odd degree, we would have a problem. Try it. It cannot be done because

> *In any graph the number of points with odd degree must be an even number.*

This is the hand-shaking theorem – the first theorem of graph theory. In any graph every line has a beginning and an end, or in other words it takes two

1930
Kuratowski proves his planar graphs theorem

1935
George Pólya develops counting techniques for graphs as algebra

1999
Eric Rains and Neil Sloane extend tree counting

people to shake hands. If we add up the degrees of every point for the whole graph we must get an even number, say N. Next we say there are x points with odd degree and y points with even degree. Adding all the degrees of the odd points together we'll have N_x and adding all the degrees of the even points will give us N_y, which is even. So we have $N_x + N_y = N$, and therefore $N_x = N - N_y$. It follows that N_x is even. But x itself cannot be odd because the addition of an odd number of odd degrees would be an odd number. So it follows that x must be even.

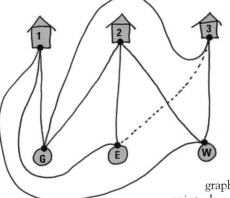

Non-planar graphs The utilities problem is an old puzzle. Imagine three houses and three utilities – gas, electricity and water. We have to connect each of the houses to each of the utilities, but there's a catch – the connections must not cross.

In fact this cannot be done – but you might try it out on your unsuspecting friends. The graph described by connecting three points to another three points in all possible ways (with only nine lines) cannot be drawn in the plane without crossings. Such a graph is called non-planar. This utilities graph, along with the graph made by all lines connecting five points, has a special place in graph theory. In 1930, the Polish mathematician Kazimierz Kuratowski proved the startling theorem that a graph is planar if and only if it does not contain either one of these two as a subgraph, a smaller graph contained within the main one.

Trees A 'tree' is a particular kind of graph, very different from the utitlities graph or the Königsberg graph. In the Königsberg bridge problem there were opportunities for starting at a point and returning to it via a different route. Such a route from a point and back to itself is called a cycle. A tree is a graph which has no cycles.

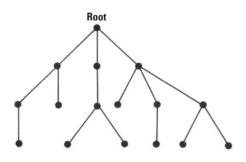

Root

A familiar example of a tree graph is the way directories are arranged in computers. They are arranged in a hierarchy with a root directory and subdirectories leading off it. Because there are no cycles there is no way to cross from one branch other than through the root directory – a familiar manoeuvre for computer users.

Counting trees How many different trees can be made from a specific number of points? The problem of counting trees was tackled by the 19th-century English mathematician Arthur Cayley. For example, there are exactly three different tree types with five points:

Cayley was able to count the number of different tree types for small numbers of points. He was able to go as far as trees with fewer than 14 points before the sheer computational complexity was too much for a man without a computer. Since then the calculations have advanced as far as trees with as many as 22 points. There are millions of possible types for these.

Even in its own time, Cayley's research had practical applications. Counting trees is relevant in chemistry, where the distinctiveness of some compounds depends on the way atoms are arranged in their molecules. Compounds with the same number of atoms but with different arrangements have different chemical properties. Using his analysis it was possible to predict the existence of chemicals 'at the tip of his pen' that were subsequently found in the laboratory.

the condensed idea
Across the bridges and into the trees

30 The four-colour problem

Who might have given young Tiny Tim a Christmas present of four coloured wax crayons and a blank county map of England? It could have been the cartographer neighbour who occasionally sent in small gifts, or that odd mathematician Augustus De Morgan, who lived nearby and passed the time of day with Tim's father. Mr Scrooge it was not.

The Cratchits lived in a drab terrace house in Bayham Street, Camden Town just north of the newly opened University College, where De Morgan was professor. The source of the gift would have become clear in the new year when the professor called to see if Tim had coloured the map.

De Morgan had definite ideas on how this should be done: 'you are to colour the map so that two counties with a common border have different colours'.

'But I haven't enough colours', said Tim without a thought. De Morgan would have smiled and left him to the task. But just recently one of his students, Frederick Guthrie, had asked him about it, and mentioned a successful colouring of England with only four colours. The problem stirred De Morgan's mathematical imagination.

Is it possible to colour *any* map with just four colours so that the regions are distinguished? Cartographers may have believed this for centuries but can it be proved rigorously? We can think of any map in the world besides the English county map, such as American states or French départements, and even artificial maps, made up of arbitrary regions and borders. Three colours, though, are not enough.

timeline

AD 1852	1879	1890
Guthrie, De Morgan's student, puts the problem to him	Kempe is believed to have solved the problem	Heawood exposes errors Kempe's proof and prove five-colour theorem

Let's look at the map of the western states of America. If only blue, green and red were available we could start off by colouring Nevada and Idaho. It does not matter which colour we begin with so we'll choose blue for Nevada and green for Idaho. So far so good. This choice would mean that Utah *must* be coloured red, and in turn Arizona green, California red, and Oregon green. This means that both Oregon and Idaho are coloured green so cannot be distinguished. But if we had four colours, with a yellow as well, we could use this to colour Oregon and everything would be satisfactory. Would these four colours – blue, green, red and yellow be sufficient for *any* map? This question is known as the four-colour problem.

The western states of America

The spread of the problem

Within 20 years of De Morgan recognizing the problem as one of significance, it became known within the mathematical community of Europe and America. In the 1860s, Charles Sanders Peirce, an American mathematician and philosopher, thought he had proved it but there is no trace of his argument.

The problem gained greater prominence through the intercession of the Victorian man of science Francis Galton. He saw publicity value in it and inveigled the eminent Cambridge mathematician Arthur Cayley to write a paper on it in 1878. Unhappily, Cayley was forced to admit that he had failed to obtain a proof, but he observed that it was sufficient to consider only cubic maps (where exactly three countries meet at a point). This contribution spurred on his student Alfred Bray Kempe to attempt a solution. Just one year later Kempe announced he had found a proof. Cayley heartily congratulated him, his proof was published, and he gained election to the Royal Society of London.

What happened next?

Kempe's proof was long and technically demanding, and though one or two people were unconvinced by it, the proof was generally accepted. There was a surprise ten years later when Durham-based Percy Heawood found an example of a map which exposed a flaw in Kempe's argument. Though he failed to find his own proof, Heawood showed that the challenge of the four-colour problem was still open. It would be back to the drawing boards for mathematicians and a chance for some new tyro to make their mark. Using some of Kempe's techniques Heawood proved a five-colour theorem – that any map could be coloured with five colours. This would have been a great result if someone could construct a map that could

1976	1994
Appel and Haken give a computer-based proof for the general result	The computer proof is simplified but remains a computer-based proof

not be coloured with four colours. As it was, mathematicians were in a quandary: was it to be four or five?

The basic four-colour problem was concerned with maps drawn on a flat or spherical surface. What about maps drawn on a surface like a donut – a surface more interesting to mathematicians for its shape than its taste. For this surface, Heawood showed that seven colours were both necessary and sufficient to colour any map drawn on it. He even proved a result for a multi-holed donut (with a number, h, holes) in which he counted the number of colours that guaranteed any map could be coloured – though he had not proved these were the minimum number of colours. A table for the first few values of Heawood's h is:

The simple donut or 'torus'

A torus with two holes

number of holes, h	1	2	3	4	5	6	7	8
sufficient number of colours, C	7	8	9	10	11	12	12	13

and in general, $C = [\frac{1}{2}(7 + \sqrt{(1 + 48h)})]$. The square brackets indicate that we only take the whole number part of the term within them. For example, when $h = 8$, $C = [13.3107...] = 13$. Heaward's formula was derived on the strict understanding that the number of holes is greater than zero. Tantalizingly the formula gives the answer $C = 4$ if the debarred value $h = 0$ is substituted.

The problem solved? After 50 years, the problem which had surfaced in 1852 remained unproved. In the 20th century the brainpower of the world's elite mathematicians was flummoxed.

Some progress was made and one mathematician proved that four colours were enough for up to 27 countries on a map, another bettered this with 31 countries and one came in with 35 countries. This nibbling process would take forever if continued. In fact the observations made by Kempe and Cayley in their very early papers provided a better way forward, and mathematicians found that they had only to check certain map configurations to guarantee that four colours were enough. The catch was that there was a large number of them – at the early stages of these attempts at proof there were thousands to check. This checking could not be done by hand but luckily the German mathematician Wolfgang Haken, who had worked on the problem for many years, was able to enlist the services of the American mathematician and computer expert Kenneth Appel. Ingenious methods lowered the number of configurations to fewer than 1500. By late June 1976, after many sleepless nights, the job was done and in partnership with their trusty IBM 370 computer, they had cracked the great problem.

The mathematics department at the University of Indiana had a new trumpet to blow. They replaced their 'largest discovered prime' postage stamp with the news that 'four colours suffice.' This was local pride but where was the general applause from the world's mathematics community? After all, this was a venerable problem which can be understood by any person of Tiny Tim's tender years, but for well over a century had teased and tortured some of the greatest mathematicians.

The applause was patchy. Some grudgingly accepted that the job had been done but many remained sceptical. The trouble was that it was a computer-based proof and this stepped right outside the traditional form of a mathematical proof. It was only to be expected that a proof would be difficult to follow, and the length could be long, but a computer proof was a step too far. It raised the issue of 'checkability'. How could anyone check the thousands of lines of computer code on which the proof depended. Errors in computer coding can surely be made. An error might prove fatal.

That was not all. What was really missing was the 'aha' factor. How could anyone read through the proof and appreciate the subtlety of the problem, or experience the crucial part of the argument, the aha moment. One of the fiercest critics was the eminent mathematician Paul Halmos. He thought that a computer proof had as much credibility as a proof by a reputable fortune teller. But many do recognize the achievement, and it would be a brave or foolish person who would spend their precious research time trying to find a counterexample of a map which required five colours. They might well have done pre Appel and Haken, but not afterwards.

After the proof Since 1976 the number of configurations to be checked has been reduced by a factor of a half and computers have become faster and more powerful. This said, the mathematical world still awaits a shorter proof along traditional lines. Meanwhile the four-colour theorem has spawned significant problems in graph theory and had the subsidiary effect of challenging the mathematician's very notion of what constitutes a mathematical proof.

the condensed idea
Four colours will be enough

31 Probability

What is the chance of it snowing tomorrow? What is the likelihood that I will catch the early train? What is the probability of you winning the lottery? Probability, likelihood, chance are all words we use every day when we want to know the answers. They are also the words of the mathematical theory of probability.

Probability theory is important. It has a bearing on uncertainty and is an essential ingredient in evaluating risk. But how can a theory involving uncertainty be quantified? After all, isn't mathematics an exact science?

The real problem is to *quantify* probability.

Suppose we take the simplest example on the planet, the tossing of a coin. What is the probability of getting a head? We might rush in and say the answer is ½ (sometimes expressed as 0.5 or 50%). Looking at the coin we make the assumption it is a fair coin, which means that the chance of getting a head equals the chance of getting a tail, and therefore the probability of a head is ½.

Situations involving coins, balls in boxes, and 'mechanical' examples are relatively straightforward. There are two main theories in the assignment of probabilities. Looking at the two-sided symmetry of the coin provides one approach. Another is the relative frequency approach, where we conduct the experiment a large number of times and count the number of heads. But how large is large? It is easy to believe that the number of heads relative to the number of tails is roughly 50:50 but it might be that this proportion would change if we continued the experiment.

But what about coming to a sensible measure of the probability of it snowing tomorrow? There will again be two outcomes: either it snows or it does not snow, but it is not at all clear that they are equally likely as it was for the coin.

timeline

C. AD 1650s	**1785**
The foundations of probability are laid by Pascal and Huygens	Condorcet applies probability to analysis of juries and electoral sy...

An evaluation of the probability of it snowing tomorrow will have to take into account the weather conditions at the time and a host of other factors. But even then it is not possible to pinpoint an exact number for this probability. Though we may not come to an actual number, we can usefully ascribe a 'degree of belief' that the probability will be low, medium or high. In mathematics, probability is measured on a scale from 0 to 1. The probability of an impossible event is 0 and a certainty is 1. A probability of 0.1 would mean a low probability while 0.9 would signify a high probability.

Origins of probability The mathematical theory of probability came to the fore in the 17th century with discussions on gambling problems between Blaise Pascal, Pierre de Fermat and Antoine Gombaud (also known as the Chevalier de Méré). They found a simple game puzzling. The Chevalier de Méré's question is this: which is more likely, rolling a 'six' on four throws of a dice, or rolling a 'double six' on 24 throws with two dice? Which option would you put your shirt on?

The prevailing wisdom of the time thought the better option was to bet on the double six because of the many more throws allowed. This view was shattered when the probabilities were analysed. Here is how the calculations go:

Throw one dice: the probability of *not* getting a six on a single throw is $\frac{5}{6}$, and in four throws the probability of this would be $\frac{5}{6} \times \frac{5}{6} \times \frac{5}{6} \times \frac{5}{6}$ which is $(\frac{5}{6})^4$. Because the results of the throws do not affect each other, they are 'independent' and we can multiply the probabilities. The probability of at least one six is therefore

$$1 - (\tfrac{5}{6})^4 = 0.517746\ldots$$

Throw two dice: the probability of *not* getting a double six in one throw is $\frac{35}{36}$ and in 24 throws this has the probability $(\frac{35}{36})^{24}$.

The probability of at least one double six is therefore

$$1 - (\tfrac{35}{36})^{24} = 0.491404\ldots$$

We can take this example a little further.

...place publishes his two volume
...nalytical Theory of Probabilities

1912
Keynes publishes his *Treatise on Probability* which influences his theories of economics and statistics

1933
Kolmogorov presents probability in an axiomatic way

Playing craps The two dice example is the basis of the modern game of craps played in casinos and online betting. When two distinguishable dice (red and blue) are thrown there are 36 possible outcomes and these may be recorded as pairs (x,y) and displayed as 36 dots against a set of x/y axes – this is called the 'sample space'.

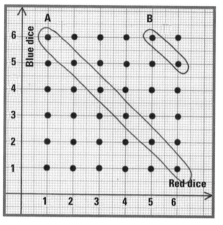

Sample space
(for 2 dice)

Let's consider the 'event' A of getting the sum of the dice to add up to 7. There are 6 combinations that each add up to 7, so we can describe the event by

$$A = \{(1,6), (2,5), (3,4), (4,3), (5,2), (6,1)\}$$

and ring it on the diagram. The probability of A is 6 chances in 36, which can be written $Pr(A) = 6/36 = 1/6$. If we let B be the event of getting the sum on the dice equal to 11 we have the event $B = \{(5,6), (6,5)\}$ and $Pr(B) = 2/36 = 1/18$.

In the dice game craps, in which two dice are thrown on a table, you can win or lose at the first stage, but for some scores all is not lost and you can go onto a second stage. You win at the first throw if either the event A or B occurs – this is called a 'natural'. The probability of a natural is obtained by adding the individual probabilities, $6/36 + 2/36 = 8/36$. You lose at the first stage if you throw a 2, 3 or a 12 (this is called 'craps'). A calculation like that above gives the probability of losing at the first stage as $4/36$. If a sum of either 4, 5, 6, 8, 9 or 10 is thrown, you go onto a second stage and the probability of doing this is $24/36 = 2/3$.

In the gaming world of casinos the probabilities are written as odds. In craps, for every 36 games you play, on average you will win at the first throw 8 times and not win 28 times so the odds against winning on the first throw are 28 to 8, which is the same as 3.5 to 1.

The monkey on a typewriter Alfred is a monkey who lives in the local zoo. He has a battered old typewriter with 26 keys for the letters of the alphabet, a key for a full stop, one for a comma, one for a question mark and one for a space – 30 keys in all. He sits in a corner filled with literary ambition, but his method of writing is curious – he hits the keys at random.

Any sequence of letters typed will have a nonzero chance of occurring, so there is a chance he will type out the plays of Shakespeare word perfect.

More than this, there is a chance (albeit smaller) he will follow this with a translation into French, and then Spanish, and then German. For good measure we could allow for the possibility of him continuing on with the poems of William Wordsworth. The chance of all this is minute, but it is certainly not zero. This is the key point. Let's see how long he will take to type the soliloquy in *Hamlet*, starting off with the opening 'To be or'. We imagine 8 boxes which will hold the 8 letters including the spaces.

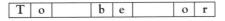

The number of possibilities for the first position is 30, for the second is 30, and so on. So the number of ways of filling out the 8 boxes is $30 \times 30 \times 30 \times 30 \times 30 \times 30 \times 30 \times 30$. The chance of Alfred getting as far as 'To be or' is 1 chance in 6.561×10^{11}. If Alfred hits the typewriter once every second there is an expectation he will have typed 'To be or' in about 20,000 years, and proved himself a particularly long-lived primate. So don't hold your breath waiting for the whole of Shakespeare. Alfred will produce nonsense like 'xo,h?yt?' for a great deal of the time.

How has the theory developed? When probability theory is applied the results can be controversial, but at least the mathematical underpinnings are reasonably secure. In 1933, Andrey Nikolaevich Kolmogorov was instrumental in defining probability on an axiomatic basis – much like the way the principles of geometry were defined two millennia before.

Probability is defined by the following axioms:

 1. the probability of all occurrences is 1
 2. probability has a value which is greater than or equal to zero
 3. when occurrences cannot coincide their probabilities can be added

From these axioms, dressed in technical language, the mathematical properties of probability can be deduced. The concept of probability can be widely applied. Much of modern life cannot do without it. Risk analysis, sport, sociology, psychology, engineering design, finance, and so on – the list is endless. Who'd have thought the gambling problems that kick-started these ideas in the 17th century would spawn such an enormous discipline? What were the chances of that?

the condensed idea
The gambler's secret system

32 Bayes's theory

The early years of the Rev. Thomas Bayes are obscure. Born in the southeast of England, probably in 1702, he became a non-conformist minister of religion, but also gained a reputation as a mathematician and was elected to the Royal Society of London in 1742. Bayes's famous _Essay towards solving a problem in the doctrine of chances_ was published in 1763, two years after his death. It gave a formula for finding inverse probability, the probability 'the other way around', and it helped create a concept central to Bayesian philosophy – conditional probability.

Thomas Bayes has given his name to the Bayesians, the adherents of a brand of statistics at variance with traditional statisticians or 'frequentists'. The frequentists adopt a view of probability based on hard numerical data. Bayesian views are centred on the famous Bayes's formula and the principle that subjective degrees of belief can be treated as mathematical probability.

Conditional probability Imagine that the dashing Dr Why has the task of diagnosing measles in his patients. The appearance of spots is an indicator used for detection but diagnosis is not straightforward. A patient may have measles without having spots and some patients may have spots without having measles. The probability that a patient has spots _given_ that they have measles is a conditional probability. Bayesians use a vertical line in their formulae to mean 'given', so if we write

prob(a patient has spots | the patient has measles)

it means the probability that a patient has spots given that they have measles. The value of _prob_(a patient has spots | the patient has measles) is not the same as _prob_(the patient has measles | the patient has spots). In relation to each other, one is the probability the other way around. Bayes's formula is the

timeline

formula of calculating one from the other. Mathematicians like nothing better than using notation to stand for things. So let's say the event of having measles is M and the event of a patient having spots is S. The symbol S̃ is the event of a patient *not* having spots and M̃ the event of *not* having measles. We can see this on a Venn diagram.

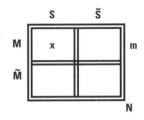

Venn diagram showing the logical structure of the appearance of spots and measles

This tells Dr Why that there are x patients who have measles and spots, m patients who have measles, while the total number of patients overall is N. From the diagram he can see that the probability that someone has measles and has spots is simply x/N, while the probability that someone has measles is m/N. The conditional probability, the probability that someone has spots given that they have measles, written $prob(S \mid M)$, is x/m. Putting these together, Dr Why gets the probability that someone has both measles and spots

$$prob(M \, \& \, S) = \frac{x}{N} = \frac{x}{m} \times \frac{m}{N}$$

or

$$prob(M \, \& \, S) = prob(S \mid M) \times prob(M)$$

and similarly

$$prob(M \, \& \, S) = prob(M \mid S) \times prob(S)$$

Bayes's formula Equating the expressions for $prob(M \, \& \, S)$ gives Bayes's formula, the relationship between the conditional probability and its inverse. Dr Why will have a good idea about $prob(S \mid M)$, the probability that if a patient has measles, they have spots. The conditional probability the other way around is what Dr Why is really interested in, his estimate of whether a patient has measles if they present with spots. Finding this out is the inverse problem and the kind of problem addressed by Bayes in his essay. To work out the probabilities we need to put in some numbers. These will be subjective but what is important is to see how they combine. The probability that if patients have measles, they have spots, $prob(S \mid M)$ will be high, say 0.9 and if the patient does not have measles, the probability of them having spots $prob(S \mid M̃)$ will be low, say 0.15. In both these situations Dr Why will have a good idea of the values of these probabilities. The dashing doctor will also have an idea about the percentage of people in the population who have measles, say 20%. This is expressed as prob(M) = 0.2. The only other piece of

$$prob(M \mid S) = \frac{prob(M)}{prob(S)} \times prob(S \mid M)$$

Bayes's formula

1950
Jimmy Savage and Dennis Lindley spearhead the modern Bayesian movement

1950s
The term 'Bayesian' comes into use for the first time

1992
The International Society for Bayesian Analysis is founded

information we need is $prob(S)$, the percentage of people in the population who have spots. Now the probability of someone having spots is the probability of someone having measles *and* spots plus the probability that someone does not have measles but does have spots. From our key relations, $prob(S) = 0.9 \times 0.2 + 0.15 \times 0.8 = 0.3$. Substituting these values into Bayes's formula gives:

$$prob(M \mid S) = \frac{0.2}{0.3} \times 0.9 = 0.6$$

The conclusion is that from all the patients with spots that the doctor sees he correctly detects measles in 60% of his cases. Suppose now that the doctor receives more information on the strain of measles so that the probability of detection goes up, that is $prob(S \mid M)$ the probability of having spots from measles, increases from 0.9 to 0.95 and $prob(S \mid \bar{M})$, the probability of spots from some other cause, declines from 0.15 to 0.1. How does this change improve his rate of measles detection? What is the new $prob(M \mid S)$? With this new information, $prob(S) = 0.95 \times 0.2 + 0.1 \times 0.8 = 0.27$, so in Bayes's formula, $prob(M \mid S)$ is 0.2 divided by $prob(S) = 0.27$ and then all multiplied by 0.95, which comes to 0.704. So Dr Why can now detect 70% of cases with this improved information. If the probabilities changed to 0.99, and 0.01 respectively then the detection probability, $prob(M \mid S)$, becomes 0.961 so his chance of a correct diagnosis in this case would be 96%.

Modern day Bayesians The traditional statistician would have little quarrel with the use of Bayes's formula where the probability can be measured. The contentious sticking point is interpreting probability as degrees of belief or, as it is sometimes defined subjective probability.

In a court of law, questions of guilt or innocence are sometimes decided by the 'balance of probabilities'. Strictly speaking this criterion only applies to civil cases but we shall imagine a scenario where it applies to criminal cases as well. The frequentist statistician has a problem ascribing any meaning to the probability of a prisoner being guilty of a crime. Not so the Bayesian who does not mind taking feelings on board. How does this work? If we are to use the balance-of-probabilities method of judging guilt and innocence we now see how the probabilities can be juggled. Here is a possible scenario.

A juror has just heard a case in court and decided that the probability of the accused being guilty is about 1 in 100. During deliberations in the jury room the jury is called back into court to hear further evidence from the prosecution. A weapon has been found in the prisoner's house and the leading prosecution counsel claims that the probability of finding it is as high as 0.95

if the prisoner is guilty, but if he is innocent the probability of finding the weapon would be only 0.1. The probability of finding a weapon in the prisoner's house is therefore much higher if the prisoner is guilty than if they are innocent. The question before the juror is how should they modify their opinion of the prisoner in the light of this new information? Using our notation again, G is the event that the prisoner is guilty and E is the event that new evidence is obtained. The juror has made an initial assessment that $prob(G) = 1/100$ or 0.01. This probability is called the prior probability. The reassessment probability $prob(G \mid E)$ is the revised probability of guilt given the new evidence E, and this is called the posterior probability. Bayes's formula in the form

$$prob(G \mid E) = \frac{prob\ (E \mid G)}{prob\ (E)} \times prob(G)$$

shows the idea of the prior probability being updated to the posterior probability $prob(G \mid E)$. Like working out $prob(S)$ in the medical example, we can work out $prob(E)$ and we find

$$prob(G \mid E) = \frac{0.95}{0.95 \times 0.01 + 0.1 \times 0.99} \times 0.01 = 0.088$$

This will present a quandary for the juror because the initial assessment of a 1% chance of guilt has risen to almost 9%. If the prosecution had made the greater claim that the probability of finding the incriminating weapon was as high as 0.99 if the prisoner is guilty but if innocent the probability of finding the weapon was only 0.01, then repeating the Bayes's formula calculation the juror would have to revise their opinion from 1% to 50%.

Using Bayes's formula in such situations has been exposed to criticisms. The leading thrust has been on how one arrives at the prior probability. In its favour Bayesian analysis presents a way of dealing with subjective probabilities and how they may be updated based on evidence. The Bayesian method has applications in areas as diverse as science, weather forecasting and criminal justice. Its proponents argue its soundness and pragmatic character in dealing with uncertainty. It has a lot going for it.

the condensed idea
Updating beliefs using evidence

33 The birthday problem

Imagine you are on the top deck of the Clapham omnibus with nothing in particular to do but count your fellow passengers going off to work in the early morning. As it is likely that all the passengers are independent of each other, we may safely assume that their birthdays are randomly scattered throughout the year. Including you there are only 23 passengers on board. It is not many, but enough to claim there is a better than even chance that two passengers share a birthday. Do you believe it? Millions do not but it is absolutely true. Even a seasoned expert in probability, William Feller, thought it astounding.

The Clapham omnibus is now too small for our needs so we resume the argument in a large room. How many people must gather in the room so that it is *certain* that two people share the same birthday? There are 365 days in a standard year (and we'll ignore leap years just to make things simpler) so if there were 366 people in the room, at least one pair would *definitely* have the same birthday. It cannot be the case that they all have different ones.

This is the pigeonhole principle: if there are $n + 1$ pigeons who occupy n pigeonholes, one hole must contain more than one pigeon. If there were 365 people we could not be certain there would be a common birthday because the birthdays could each be on different days of the year. However, if you take 365 people at random this would be extremely unlikely and the probability of two people not sharing a birthday would be minuscule. Even if there are only 50 people in the room there is a 96.5% chance that two people share a birthday.

timeline

AD **1654**	**1657**	**1718**
Blaise Pascal lays the foundations of probability theory	Christiaan Huygens writes the first published work on probability	Abraham de Moivre publishes *The Doctrine of Chance*, with expanded editions following in 1738 and 1756

If the number of people is reduced still further the probability of two sharing a birthday reduces. We find that 23 people is the number for which the probability is just greater than ½ and for 22 people the probability that a birthday is shared is just less than ½. The number 23 is the critical value. While the answer to the classic birthday problem is surprising it is not a paradox.

Can we prove it? How can we be convinced? Let's select a person at random. The probability that another person has the same birthday as this person is 1/365 and so the probability these two do not share a birthday is one minus this (or 364/365). The probability that yet another person selected at random shares a birthday with the first two is 2/365 so the probability this person does not share a birthday with either of the first two is one minus this (or 363/365). The probability of none of these three sharing a birthday is the multiplication of these two probabilities, or (364/365) × (363/365) which is 0.9918.

Continuing this line of thought for 4, 5, 6, . . . people unravels the birthday problem paradox. When we get as far as 23 people with our pocket calculator we get the answer 0.4927 as the probability that none of them shares a birthday. The negation of 'none of them sharing a birthday' is 'at least two people share a birthday' and the probability of this is 1 − 0.4927 = 0.5073, just greater than the crucial ½.

If $n = 22$, the probability of two people sharing a birthday is 0.4757, which is less than ½. The apparent paradoxical nature of the birthday problem is bound up with language. The birthday result makes a statement about two people sharing a birthday, but it does not tell us which two people they are. We do not know where the matches will fall. If Mr Trevor Thomson whose birthday is on 8 March is in the room, a different question might be asked.

How many birthdays coincide with Mr Thomson's? For this question, the calculation is different. The probability of Mr Thomson not sharing his birthday with another person is 364/365 so that the probability that he does not share his birthday with any of the other $n − 1$ people in the room is $(364/365)^{n-1}$. Therefore the probability that Mr Thomson does share his birthday with someone will be one minus this value.

1920s

Bose considers Einstein's theory of light as an occupancy problem

1939

Richard von Mises proposes the birthday problem

If we compute this for $n = 23$ this probability is only 0.061151 so there is only a 6% chance that someone else will have their birthday on 8 March, the same date as Mr Thomson's birthday. If we increase the value of n, this probability will increase. But we have to go as far $n = 254$ (which includes Mr Thomson in the count) for the probability to be greater than ½. For $n = 254$, its value is 0.5005. This is the cutoff point because $n = 253$ will give the value 0.4991 which is less than ½. There will have to be a gathering of 254 people in the room for a chance greater than ½ that Mr Thomson shares his birthday with someone else. This is perhaps more in tune with our intuition than with the startling solution of the classic birthday problem.

Other birthday problems The birthday problem has been generalized in many ways. One approach is to consider three people sharing a birthday. In this case 88 people would be required before there is a better than even chance that three people will share the same birthday. There are correspondingly larger groups if four people, five people, . . . are required to share a birthday. In a gathering of 1000 people, for example, there is a better than even chance that nine of them share a birthday.

Other forays into the birthday problem have inquired into near birthdays. In this problem a match is considered to have occurred if one birthday is within a certain number of days of another birthday. It turns out that a mere 14 people in a room will give a greater than even chance of two people having a birthday in common or having a birthday within a day of each other.

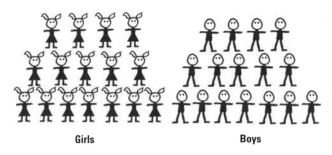

Girls   Boys

A variant of the birthday problem which requires more sophisticated mathematical tools is the birthday problem for boys and girls: if a class consists of an equal number of boys and girls, what would be the minimum group that would give a better than even chance that a boy and a girl shared a birthday?

The result is that a class of 32 (16 girls and 16 boys) would yield the minimum group. This can be compared with 23 in the classic birthday problem.

By changing the question slightly we can get other novelties (but they are not easy to answer). Suppose we have a long queue forming outside a Bob Dylan concert and people join it randomly. As we are interested in birthdays we may discount the possibility of twins or triplets arriving together. As the fans enter they are asked their birthdays. The mathematical question is this: how many people would you expect to be admitted before two *consecutive* people have the same birthday? Another question: How many people go into the concert hall before a person turns up with the same birthday as Mr Trevor Thomson (8 March)?

The birthday calculation makes the assumption that birthdays are uniformly distributed and that each birthday has an equal chance of occurring for a person selected at random. Experimental results show this is not exactly true (more are born during the summer months) but it is close enough for the solution to be applicable.

Birthday problems are examples of occupancy problems, in which mathematicians think about placing balls in cells. In the birthday problem, the number of cells is 365 (these are identified with possible birthdays) and the balls to be placed at random in the cells are the people. The problem can be simplified to investigate the probability of two balls falling in the same cell. For the boys-and-girls problem, the balls are of two colours.

It is not only mathematicians who are interested in the birthday problem. Satyendra Nath Bose was attracted to Albert Einstein's theory of light based on photons. He stepped out of the traditional lines of research and considered the physical setup in terms of an occupancy problem. For him, the cells were not days of the year as in the birthday problem but energy levels of the photons. Instead of people being put into cells as in the birthday problem he distributed numbers of photons. There are many applications of occupancy problems in other sciences. In biology, for instance, the spread of epidemics can be modelled as an occupancy problem – the cells in this case are geographical areas, the balls are diseases and the problem is to figure out how the diseases are clustered.

The world is full of amazing coincidences but only mathematics gives us the way of calculating their probability. The classical birthday problem is just the tip of the iceberg in this respect and it is a great entry into serious mathematics with important applications.

the condensed idea
Calculating coincidences

34 Distributions

Ladislaus J. Bortkiewicz was fascinated by mortality tables. Not for him a gloomy topic, they were a field of enduring scientific enquiry. He famously counted the number of cavalrymen in the Prussian army that had been killed by horse-kicks. Then there was Frank Benford, an electrical engineer who counted the first digits of different types of numerical data to see how many were ones, twos and so on. And George Kingsley Zipf, who taught German at Harvard, had an interest in philology and analysed the occurrences of words in pieces of text.

All these examples involve measuring the probabilities of events. What are the probabilities of x cavalrymen in a year receiving a lethal kick from a horse? Listing the probabilities for each value of x is called a distribution of probabilities, or for short, a probability distribution. It is also a *discrete* distribution because the values of x only take isolated values – there are gaps between the values of interest. You can have three or four Prussian cavalrymen struck down by a lethal horse-kick but not 3½. As we'll see, in the case of the Benford distribution we are only interested in the appearance of digits 1, 2, 3, . . . and, for the Zipf distribution, you may have the word 'it' ranked eighth in the list of leading words, but not at position, say 8.23.

Life and death in the Prussian army Bortkiewicz collected records for ten corps over a 20 year period giving him data for 200 corps-years. He looked at the number of deaths (this was what mathematicians call the variable) and the number of corps-years when this number of deaths occurred. For example, there were 109 corps-years when no deaths occurred, while in one corps-year, there were four deaths. At the barracks, Corp C (say) in one particular year experienced four deaths.

timeline

How is the number of deaths distributed? Collecting this information is one side of the statistician's job – being out in the field recording results. Bortkiewicz obtained the following data:

Number of deaths	0	1	2	3	4
Frequency	109	65	22	3	1

Thankfully, being killed by a horse-kick is a rare event. The most suitable theoretical technique for modelling how often rare events occur is to use something called the Poisson distribution. With this technique, could Bortkiewicz have predicted the results without visiting the stables? The theoretical Poisson distribution says that the probability that the number of deaths (which we'll call X) has the value x is given by the Poisson formula, where e is the special number discussed earlier that's associated with growth (see page 24) and the exclamation mark means the factorial, the number multiplied by all the other whole numbers between it and 1 (see page 26). The Greek letter lambda, written λ, is the average number of deaths. We need to find this average over our 200 corps-years so we multiply 0 deaths by 109 corps-years (giving 0), 1 death by 65 corps-years (giving 65), 2 deaths by 22 corps-years (giving 44), 3 deaths by 3 corps-years (giving 9) and 4 deaths by 1 corps-year (giving 4) and then we add all of these together (giving 122) and divide by 200. So our average number of deaths per corps-year is 122/200 = 0.61.

$$e^{-\lambda}\lambda^x/x\,!$$

The Poisson formula

The theoretical probabilities (which we'll call p) can be found by substituting the values r = 0, 1, 2, 3 and 4 into the Poisson formula. The results are:

Number of deaths	0	1	2	3	4
Probabilities, p	0.543	0.331	0.101	0.020	0.003
Expected number of deaths, 200 × p	108.6	66.2	20.2	4.0	0.6

It looks as though the theoretical distribution is a good fit for the experimental data gathered by Bortkiewicz.

First numbers If we analyse the last digits of telephone numbers in a column of the telephone directory we would expect to find 0, 1, 2, . . . , 9 to be uniformly distributed. They appear at random and any number has an equal chance of turning up. In 1938 the electrical engineer Frank Benford found that

1938	1950	2003
Benford restates the law of distribution of first digits	Zipf derives a formula relating word use to vocabulary	The Poisson distribution is used in the analysis of fish stocks in the North Atlantic

this was not true for the first digits of some sets of data. In fact he rediscovered a law first observed by the astronomer Simon Newcomb in 1881.

Yesterday I conducted a little experiment. I looked through the foreign currency exchange data in a national newspaper. There were exchange rates like 2.119 to mean you will need (US dollar) $2.119 to buy £1 sterling. Likewise, you will need (Euro) €1.59 to buy £1 sterling and (Hong Kong dollar) HK $15.390 to buy £1. Reviewing the results of the data and recording the number of appearances by first digit, gave the following table:

First digit	1	2	3	4	5	6	7	8	9	Total
Number of occurrences	18	10	3	1	3	5	7	2	1	**50**
Percentage, %	36	20	6	2	6	10	14	4	2	**100**

These results support Benford's law, which says that for some classes of data, the number 1 appears as the first digit in about 30% of the data, the number 2 in 18% of the data and so on. It is certainly not the uniform distribution that occurs in the last digit of the telephone numbers.

It is not obvious why so many data sets do follow Benford's law. In the 19th century when Simon Newcomb observed it in the use of mathematical tables he could hardly have guessed it would be so widespread.

Instances where Benford's distribution can be detected include scores in sporting events, stock market data, house numbers, populations of countries, and the lengths of rivers. The measurement units are unimportant – it does not matter if the lengths of rivers are measured in metres or miles. Benford's law has practical applications. Once it was recognized that accounting information followed this law, it became easier to detect false information and uncover fraud.

Words   One of G.K. Zipf's wide interests was the unusual practice of counting words. It turns out that the ten most popular words appearing in the English language are the tiny words ranked as shown:

Rank	1	2	3	4	5	6	7	8	9	10
Word	the	of	and	to	a	in	that	it	is	was

This was found by taking a large sample across a wide range of written work and just counting words. The most common word was given rank 1, the next rank 2, and so on. There might be small differences in the popularity stakes if a range of texts were analysed, but it will not vary much.

It is not surprising that 'the' is the most common, and 'of' is second. The list continues and you might want to know that 'among' is in 500th position and 'neck' is ranked 1000. We shall only consider the top ten words. If you pick up a text at random and count these words you will get more or less the same words in rank order. The surprising fact is that the ranks have a bearing on the actual number of appearances of the words in a text. The word 'the' will occur twice as often as 'of' and three times more frequently than 'and', and so on. The actual number is given by a well-known formula. This is an experimental law and was discovered by Zipf from data. The theoretical Zipf's law says that the percentage of occurrences of the word ranked r is given by

$$\frac{k}{r} \times 100$$

where the number k depends only on the size of the author's vocabulary. If an author had command of all the words in the English language, of which there are around a million by some estimates, the value of k would be about 0.0694. In the formula for Zipf's law the word 'the' would then account for about 6.94% of all words in a text. In the same way 'of' would account for half of this, or about 3.47% of the words. An essay of 3000 words by such a talented author would therefore contain 208 appearances of 'the' and 104 appearances of the word 'of'.

For writers with only 20,000 words at their command, the value of k rises to 0.0954, so there would be 286 appearances of 'the' and 143 appearances of the word 'of'. The smaller the vocabulary, the more often you will see 'the' appearing.

Crystal ball gazing Whether Poisson, Benford or Zipf, all these distributions allow us to make predictions. We may not be able to predict a dead cert but knowing how the probabilities distribute themselves is much better than taking a shot in the dark. Add to these three, other distributions like the binomial, the negative binomial, the geometric, the hypergeometric, and many more, the statistician has an effective array of tools for analysing a vast range of human activity.

the condensed idea
Predicting how many

35 The normal curve

The 'normal' curve plays a pivotal role in statistics. It has been called the equivalent of the straight line in mathematics. It certainly has important mathematical properties but if we set to work analysing a block of raw data we would rarely find that it followed a normal curve exactly.

The normal curve is prescribed by a specific mathematical formula which creates a bell-shaped curve; a curve with one hump and which tails away on either side. The significance of the normal curve lies less in nature and more in theory, and in this it has a long pedigree. In 1733 Abraham de Moivre, a French Huguenot who fled to England to escape religious persecution, introduced it in connection with his analysis of chance. Pierre Simon Laplace published results about it and Carl Friedrich Gauss used it in astronomy, where it is sometimes referred to as the Gaussian law of error.

Adolphe Quetelet used the normal curve in his sociological studies published in 1835, in which he measured the divergence from the 'average man' by the normal curve. In other experiments he measured the heights of French conscripts and the chest measurements of Scottish soldiers and assumed these followed the normal curve. In those days there was a strong belief that most phenomena were 'normal' in this sense.

The cocktail party Let's suppose that Georgina went to a cocktail party and the host, Sebastian, asked her if she had come far? She realized afterwards it was a very useful question for cocktail parties – it applies to everyone and invites a response. It is not taxing and it starts the ball rolling if conversation is difficult.

timeline

The next day, slightly hungover, Georgina travelled to the office wondering if her colleagues had come far to work. In the staff canteen she learned that some lived around the corner and some lived 50 miles away – there was a great deal of variability. She took advantage of the fact that she was the Human Resources Manager of a very large company to tack a question on the end of her annual employee questionnaire: 'how far have you travelled to work today?' She wanted to work out the average distance of travel of the company's staff. When Georgina drew a histogram of results the distribution showed no particular form, but at least she could calculate the average distance travelled.

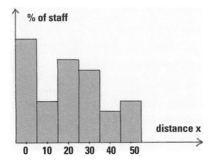

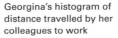

Georgina's histogram of distance travelled by her colleagues to work

This average turned out to be 20 miles. Mathematicians denote this by the Greek letter mu, written μ, and so here $\mu = 20$. The variability in the population is denoted by the Greek letter sigma, written σ, which is sometimes called the standard deviation. If the standard deviation is small the data is close together and has little variability, but if it is large, the data is spread out. The company's marketing analyst, who had trained as a statistician, showed Georgina that she might have got around the same value of 20 by sampling. There was no need to ask *all* the employees. This estimation technique depends on the Central Limit Theorem.

Take a random sample of staff from all of the company's workforce. The larger the sample the better, but 30 employees will do nicely. In selecting this sample at random it is likely there will be people who live around the corner and some long-distance travellers as well. When we calculate the average distance for our sample, the effect of the longer distances will average out the shorter distances. Mathematicians write the average of the sample as $\bar{x}$, which is read

1835	1870s	1901
Quetelet uses the normal curve to measure divergence from the average man	The distribution acquires the name 'normal'	Aleksandr Lyapunov proves the Central Limit Theorem rigorously using characteristic functions

as 'x bar'. In Georgina's case, it is most likely that the value of $\bar{x}$ will be near 20, the average of the population. Though it is certainly possible, it is unlikely that the average of the sample will be very small or very large.

The Central Limit Theorem is one reason why the normal curve is important to statisticians. It states that the actual distribution of the sample averages $\bar{x}$ approximates to a normal curve whatever the distribution of x. What does this mean? In Georgina's case, x represents the distance from the workplace and $\bar{x}$ is the average of a sample. The distribution of x in Georgina's histogram is nothing like a bell-shaped curve, but the distribution of $\bar{x}$ is, and it is centred on $\mu = 20$.

20   **Average distance x̄**

How the sample average is distributed

This is why we can use the average of a sample $\bar{x}$ as an estimate of the population average μ. The variability of the sample averages $\bar{x}$ is an added bonus. If the variability of the x values is the standard deviation σ, the variability of $\bar{x}$ is $\sigma/\sqrt{n}$ where n is the size of the sample we select. The larger the sample size, the narrower will be the normal curve, and the better will be the estimate of μ.

Other normal curves Let's do a simple experiment. We'll toss a coin four times. The chance of throwing a head each time is $p = \frac{1}{2}$. The result for the four throws can be recorded using H for heads and T for tails, arranged in the order in which they occur. Altogether there are 16 possible outcomes. For example, we might obtain three heads in the outcome THHH. There are in fact four possible outcomes giving three heads (the others are HTHH, HHTH, HHHT) so the probability of three heads is 4/16 = 0.25.

With a small number of throws, the probabilities are easily calculated and placed in a table, and we can also calculate how the probabilities are distributed. The number of combinations row can be found from Pascal's triangle (see page 52):

Number of heads	0	1	2	3	4
Number of combinations	1	4	6	4	1
Probability	0.0625	0.25	0.375	0.25	0.0625
	(= 1/16)	(= 4/16)	(= 6/16)	(= 4/16)	(= 1/16)

This is called a binomial distribution of probabilities, which occurs where there are two possible outcomes (here a head or a tail). These probabilities may be represented by a diagram in which both the heights *and* areas describe them.

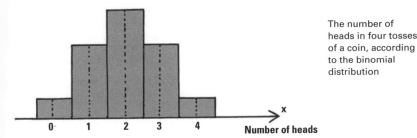

The number of heads in four tosses of a coin, according to the binomial distribution

Tossing the coin four times is a bit restrictive. What happens if we throw it a large number, say 100, times? The binomial distribution of probabilities can be applied where $n = 100$, but it can usefully be approximated by the normal bell-shaped curve with mean $\mu = 50$ (as we would expect 50 heads when tossing a coin 100 times) and variability (standard deviation) of $\sigma = 5$. This is what de Moivre discovered in the 16th century.

For large values of n, the variable x which measures the number of successes fits the normal curve increasingly well. The larger the value of n the better the approximation and tossing the coin 100 times qualifies as large. Now let's say we want to know the probability of throwing between 40 and 60 heads. The area A shows the region we're interested in and gives us the probability of tossing between 40 and 60 heads which we write as $prob(40 \leq x \leq 60)$. To find the actual numerical value we need to use pre-calculated mathematical tables, and once this has been done, we find $prob(40 \leq x \leq 60) = 0.9545$. This shows that getting between 40 and 60 heads in 100 tosses of a coin is 95.45%, which means that this is very likely.

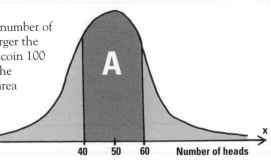

Distribution of the probability for the number of heads in 100 throws of a coin

The area left over is $1 - 0.9545$ which is a mere 0.0455. As the normal curve is symmetric about its middle, half of this will give the probability of getting more than 60 heads in a 100 tosses of the coin. This is just 2.275% and represents a very slim chance indeed. If you visit Las Vegas this would be a bet to leave well alone.

the condensed idea
The ubiquitous
bell-shaped curve

36 Connecting data

How are two sets of data connected? Statisticians of a hundred years ago thought they had the answer. Correlation and regression go together like a horse and carriage, but like this pairing, they are different and have their own jobs to do. Correlation measures how well two quantities such as weight and height are related to each other. Regression can be used to predict the values of one property (say weight) from the other (in this case, height).

Pearson's correlation The term correlation was introduced by Francis Galton in the 1880s. He originally termed it 'co-relation', a better word for explaining its meaning. Galton, a Victorian gentleman of science, had a desire to measure everything and applied correlation to his investigations into pairs of variables: the wing length and tail length of birds, for instance. The Pearson correlation coefficient, named after Galton's biographer and protégé Karl Pearson, is measured on a scale between minus one and plus one. If its numerical value is high, say +0.9, there is said to be a strong correlation between the variables. The correlation coefficient measures the tendency for data to lie along a straight line. If it is near to zero the correlation is practically non-existent.

We frequently wish to work out the correlation between two variables to see how strongly they are connected. Let's take the example of the sales of sunglasses and see how this relates to the sales of ice creams. San Francisco would be a good place in which to conduct our study and we shall gather data each month in that city. If we plot points on a graph where the x (horizontal) coordinate represents sales of sunglasses and the y (vertical) coordinate gives the sales of ice creams, each month we will have a data point (x, y) representing both pieces of data. For example, the point $(3, 4)$ could mean the May sales of sunglasses were $30,000 while sales of ice creams in the city were $40,000 in that same month. We can plot the monthly data points (x, y) for a whole year on a scatter diagram. For this example, the value of the Pearson

timeline

AD1806	1809	1885–8
Adrien-Marie Legendre fits data by least squares	Carl Friedrich Gauss uses the least-squares method in astronomical problems	Galton introduces regression and correlat

correlation coefficient would be around +0.9 indicating a strong correlation. The data has a tendency to follow a straight line. It is positive because the straight line has a positive gradient – it is pointing in a northeasterly direction.

Cause and correlation Finding a strong correlation between two variables is not sufficient to claim that one causes the other. There may be a cause and effect relation between the two variables but this cannot be claimed on the basis of numerical evidence alone. On the cause/correlation issue it is customary to use the word 'association' and wise to be wary of claiming more than this.

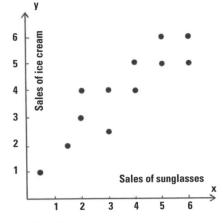

Scatter diagram

In the sunglasses and ice cream example, there is a strong correlation between the sales of sunglasses and that of ice cream. As the sales of sunglasses increase, the number of ice creams sold tends to increase. It would be ludicrous to claim that the expenditure on sunglasses *caused* more ice creams to be sold. With correlation there may be a hidden intermediary variable at work. For example, the expenditure on sunglasses and on ice creams is linked together as a result of seasonal effects (hot weather in the summer months, cool weather in the winter). There is another danger in using correlation. There may be a high correlation between variables but no logical or scientific connection at all. There could be a high correlation between house numbers and the combined ages of the house's occupants but reading any significance into this would be unfortunate.

Spearman's correlation Correlation can be put to other uses. The correlation coefficient can be adapted to treat ordered data – data where we want to know first, second, third, and so on, but not necessarily other numerical values.

Occasionally we have only the ranks as data. Let's look at Albert and Zac, two strong-minded ice skating judges at a competition who have to evaluate skaters on artistic merit. It will be a subjective evaluation. Albert and Zac have both won Olympic medals and are called on to judge the final group which has been narrowed down to five competitors: Ann, Beth, Charlotte, Dorothy and Ellie. If Albert and Zac ranked them in exactly the same way, that would be fine but life is not like that. On the other hand we would not expect Albert to rank them in one way and Zac to rank them in the very reverse order. The reality is that the rankings would be in between these two extremes. Albert ranked them 1 to 5 with Ann (the best) followed by Ellie, Beth, Charlotte and finally Dorothy in 5th position. Zac rated Ellie the best, followed by Beth, Ann, Dorothy and Charlotte. These rankings can be summarized in a table.

Skater	Albert's rankings	Zac's rankings	Difference in ranks, d	d^2
Ann	1	3	-2	4
Ellie	2	1	1	1
Beth	3	2	1	1
Charlotte	4	5	-1	1
Dorothy	5	4	1	1
$n = 5$			Sum	8

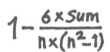

$$1 - \frac{6 \times Sum}{n \times (n^2 - 1)}$$

Spearman's formula

How can we measure the level of agreement between the judges? Spearman's correlation coefficient is the instrument mathematicians use to do this for ordered data. Its value here is +0.6 which indicates a limited measure of agreement between Albert and Zac. If we treat the pairs of ranks as points we can plot them on a graph to obtain a visual representation of how closely the two judges agree.

The formula for this correlation coefficient was developed in 1904 by the psychologist Charles Spearman who, like Pearson, was influenced by Francis Galton.

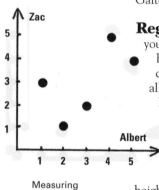

Measuring agreement between two judges

Regression lines Are you shorter or taller than both your parents or do you fall between their heights? If we were all taller than our parents, and this happened at each generation, then one day the population might be composed of ten-footers and upwards, and surely this cannot be. If we were all shorter than our parents then the population would gradually diminish in height and this is equally unlikely. The truth lies elsewhere.

Francis Galton conducted experiments in the 1880s in which he compared the heights of mature young adults with the heights of their parents. For each value of the x variable measuring parents' height (actually combining height of mother and father into a 'mid-parent' height) he observed the heights of their offspring. We are talking about a practical scientist here, so out came the pencils and sheets of paper divided into squares on which he plotted the data. For 205 mid-parents and 928 offspring he found the average height of both sets to be 68¼ inches or 5 feet 8¼ inches (173.4 cm) which value he called the mediocrity. He found that children of very tall mid-parents were generally taller than this mediocrity but not as tall as their mid-parents, while shorter children were taller than their mid-parents but shorter than the mediocrity. In other words, the children's heights regressed towards the mediocrity. It's a bit like top class batter Alex Rodriguez's performances for the New York Yankees. His batting average in an exceptional season is likely to be followed by an inferior average in the next, yet overall would still be better than the average for all players in the league. We say his batting average has regressed to the average (or mean).

Regression is a powerful technique and is widely applicable. Let's suppose that, for a survey, the operational research team of a popular retail chain chooses five of its stores, from small outlets (with 1000 customers a month) through to mega-stores (with 10,000 customers a month). The research team observes the number of staff employed in each. They plan to use regression to estimate how many staff they will need for their other stores.

Number of customers (1000s)	1	4	6	9	10
Number of staff	24	30	46	47	53

Let's plot this on a graph, where we'll make the x coordinate the number of customers (we call this the explanatory variable) while the number of staff is plotted as the y coordinate (called the response variable). It is the number of customers that explains the number of staff needed and not the other way around. The average number of customers in the stores is plotted as 6 (i.e. 6000 customers) and the average number of staff in the stores is 40. The regression line always passes through the 'average point', here (6, 40). There are formulae for calculating the regression line, the line which best fits the data (also known as the line of least squares). In our case the line is $\hat{y} = 20.8 + 3.2x$ so the slope is 3.2 and is positive (going up from left to right). The line crosses the vertical y axis at the point 20.8. The term $\hat{y}$ is the estimate of the y value obtained from the line. So if we want to know how many staff should be employed in a store that receives 5000 customers a month we could substitute the value $x = 5$ into the regression equation and obtain the estimate $\hat{y} = 37$ staff showing how regression has a very practical purpose.

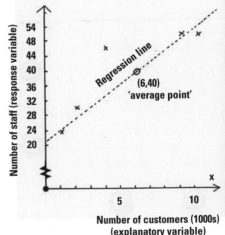

Number of customers (1000s) (explanatory variable)

the condensed idea
The interaction of data

37 Genetics

Genetics is a branch of biology, so why is it in a mathematics book? The answer is that these two subjects cross-fertilize and enrich each other. The problems of genetics require mathematics but genetics has also suggested new branches of algebra. Gregor Mendel is central to the whole theme of genetics, the study of human inheritance. Hereditary characteristics such as eye colour, hair colour, colour-blindness, left/right-handedness and blood group types are all determined by factors (alleles) of a gene. Mendel said that these factors pass independently into the next generation.

So how could eye-colour factor be transmitted to the next generation? In the basic model there are two factors, b and B:

b is the blue eyes factor
B is the brown eyes factor

In individuals, the factors appear in pairs giving rise to possible genotypes bb, bB and BB (because bB is the same as Bb). A person carries one of these three genotypes, which determines their eye colour. For example, a population could consist of a fifth of people with the genotype bb, another fifth with the genotype bB and the remaining three-fifths with the genotype BB. In terms of percentages, these genotypes would make up 20%, 20% and 60% of the population. This can be represented by a diagram showing these proportions of genotypes.

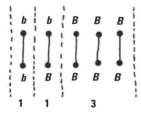

Population representing the proportions 1:1:3 of genotypes bb, bB and BB

The factor B, which denotes brown eye colour, is the dominant factor and b, the blue eye colour is the recessive factor. A person with a pure genes genotype BB will have brown eyes, but so too will a person with mixed factors, that is, those with a hybrid genotype bB because B is dominant. A person with the pure genes genotype bb will be the only genotype to show blue eyes.

timeline

AD**1718**

Abraham de Moivre publishes
the *Doctrine of Chances*

1865

Mendel proposes the existence
of genes and laws of inheritance

A burning question in the field of biology arose at the beginning of the 19th century. Would brown eyes eventually take over and blue eyes die out? Would blue eyes become extinct? The answer was a resounding 'No'.

The Hardy–Weinberg law
This was explained by the Hardy–Weinberg law, an application of basic mathematics to genetics. It explains how, in the Mendelian theory of inheritance, a dominant gene does *not* take over completely and a recessive gene does not die out.

G.H. Hardy was an English mathematician who prided himself on the non-applicability of mathematics. He was a great researcher in pure mathematics but is probably more widely known for this single contribution to genetics – which started life as a piece of mathematics on the back of an envelope done after a cricket match. Wilhelm Weinberg came from a very different background. A general medical practitioner in Germany, he was a geneticist all his life. He discovered the law at the same time as Hardy, around 1908.

The law relates to a large population in which mating happens at random. There are no preferred pairings so that, for instance, blue-eyed people do not prefer to mate with blue-eyed people. After mating, the child receives one factor from each parent. For example, a hybrid genotype bB mating with a hybrid bB can produce any one of bb, bB, BB, but a bb mating with a BB can only produce a hybrid bB. What is the probability of a b-factor being transmitted? Counting the number of b-factors there are two b-factors for each bb genotype and one b factor for each bB genotype giving, as a proportion, a total of three b-factors out of 10 (in our example of a population with 1:1:3 proportions of the three genotypes). The transmission probability of a b-factor being included in the genotype of a child is therefore 3/10 or 0.3. The transmission probability of a B-factor being included is 7/10 or 0.7. The probability of the genotype bb being included in the next generation, for example, is therefore $0.3 \times 0.3 = 0.09$. The complete set of probabilities is summarized in the table.

	b		B	
b	bb	$0.3 \times 0.3 = 0.09$	bB	$0.3 \times 0.7 = 0.21$
B	Bb	$0.3 \times 0.7 = 0.21$	BB	$0.7 \times 0.7 = 0.49$

The hybrid genotypes bB and Bb are identical so the probability of this occurring is $0.21 + 0.21 = 0.42$. Expressed as percentages, the ratios of

1908
Hardy and Weinberg show why dominant genes do not supplant recessive genes

1918
Fisher reconciles Darwin's theory with the Mendelian theory of heredity

1953
The double helix structure of DNA is discovered

genotypes *bb*, *bB* and *BB* in the new generation are 9%, 42% and 49%. Because *B* is the dominant factor, 42% + 49% = 91% of the first generation will have brown eyes. Only an individual with genotype *bb* will display the observable characteristics of the *b* factor, so only 9% of the population will have blue eyes.

The initial distribution of genotypes was 20%, 20% and 60% and in the new generation the distribution of genotypes is 9%, 42% and 49%. What happens next? Let's see what happens if a new generation is obtained from this one by random mating. The proportion of *b*-factors is 0.09 + ½ × 0.42 = 0.3, the proportion of *B*-factors is ½ × 0.42 + 0.49 = 0.7. These are *identical* to the previous transmission probabilities of the factors *b* and *B*. The distribution of genotypes *bb*, *bB* and *BB* in the further generation is therefore the same as for the previous generation, and in particular the genotype *bb* which gives blue eyes does not die out but remains stable at 9% of the population. Successive proportions of genotypes during a sequence of random matings are therefore

$$20\%, 20\%, 60\% \rightarrow 9\%, 42\%, 49\% \rightarrow \ldots \rightarrow 9\%, 42\%, 49\%$$

This is in accordance with the Hardy–Weinberg law: after one generation the genotype proportions remain constant from generation to generation, and the transmission probabilities are constant too.

Hardy's argument To see that the Hardy–Weinberg law works for *any* initial population, not just the 20%, 20% and 60% one that we selected in our example, we can do no better than refer to Hardy's own argument which he wrote to the editor of the American journal *Science* in 1908.

Hardy begins with the initial distribution of genotypes *bb*, *bB* and *BB* as *p*, *2r* and *q* and the transmission probabilities *p* + *r* and *r* + *q*. In our numerical example (of 20%, 20%, 60%), *p* = 0.2, *2r* = 0.2 and *q* = 0.6. The transmission probabilities of the factors *b* and *B* are *p* + *r* = 0.2 + 0.1 = 0.3 and *r* + *q* = 0.1 + 0.6 = 0.7. What if there were a different initial distribution of the genotypes *bb*, *bB* and *BB* and we started with, say, 10%, 60% and 30%? How would the Hardy–Weinberg law work in this case? Here we would have *p* = 0.1, *2r* = 0.6 and *q* = 0.3 and the transmission probabilities of the factors *b* and *B* are respectively *p* + *r* = 0.4 and *r* + *q* = 0.6. So the distribution of next generation of genotypes is 16%, 48% and 36%. Successive proportions of the genotypes *bb*, *bB*, and *BB* after random matings are

$$10\%, 60\%, 30\% \rightarrow 16\%, 48\%, 36\% \rightarrow \ldots \rightarrow 16\%, 48\%, 36\%$$

and the proportions settles down after one generation, as before, and the transmission probabilities of 0.4 and 0.6 remain constant. With these figures 16% of the population will have blue eyes and 48% + 36% = 84% will have brown eyes because B is dominant in the genotype bB.

So the Hardy–Weinberg law implies that these proportions of genotypes bb, bB and BB will remain constant from generation to generation whatever the initial distribution of factors in the population. The dominant B gene does not take over and the proportions of genotypes are intrinsically stable.

Hardy stressed that his model was only approximate. Its simplicity and elegance depended on many assumptions which do not hold in real life. In the model the probability of gene mutation or changes in the genes themselves has been discounted, and the consequence of the transmission proportions being constant means it has nothing to say about evolution. In real life there is 'genetic drift' and the transmission probabilities of the factors do not stay constant. This will cause variations in the overall proportions and new species will evolve.

The Hardy–Weinberg law drew together Mendel's theory – the 'quantum theory' of genetics – and Darwinism and natural selection in an intrinsic way. It awaited the genius of R.A. Fisher to reconcile the Mendelian theory of inheritance with the continuous theory where characteristics evolve.

What was missing in the science of genetics until the 1950s was a physical understanding of the genetic material itself. Then there was a dramatic advance contributed by Francis Crick, James Watson, Maurice Wilkins and Rosalind Franklin. The medium was deoxyribonucleic acid or DNA. Mathematics is needed to model the famous double helix (or a pair of spirals wrapped around a cylinder). The genes are located on segments of this double helix.

Mathematics is indispensable in studying genetics. From the basic geometry of the spirals of DNA and the potentially sophisticated Hardy–Weinberg law, mathematical models dealing with many characteristics (not just eye-colour) including male–female differences and also non-random mating have been developed. The science of genetics has also repaid the compliment to mathematics by suggesting new branches of abstract algebra of interest for their intriguing mathematical properties.

the condensed idea
Uncertainty in the gene pool

38 Groups

Evariste Galois died in a duel aged 20, but left behind enough ideas to keep mathematicians busy for centuries. These involved the theory of groups, mathematical constructs that can be used to quantify symmetry. Apart from its artistic appeal, symmetry is the essential ingredient for scientists who dream of a future theory of everything. Group theory is the glue which binds the 'everything' together.

Symmetry is all around us. Greek vases have it, snow crystals have it, buildings often have it and some letters of our alphabet have it. There are several sorts of symmetry: chief among them are mirror symmetry and rotational symmetry. We'll just look at two-dimensional symmetry – all our objects of study live on the flat surface of this page.

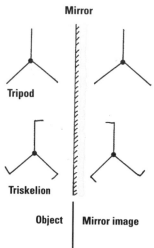

Tripod

Triskelion

Object   Mirror image

Mirror symmetry Can we set up a mirror so that an object looks the same in front of the mirror as in the mirror? The word MUM has mirror symmetry, but HAM does not; MUM in front of the mirror is the same as MUM in the mirror while HAM becomes MAH. A tripod has mirror symmetry, but the triskelion (tripod with feet) does not. The triskelion as the object before the mirror is right-handed but its mirror image in what is called the image plane is left-handed.

Rotational symmetry We can also ask whether there is an axis perpendicular to the page so that the object can be rotated in the page through an angle and be brought back to its original position. Both the tripod and the triskelion have rotational symmetry. The triskelion, meaning 'three legs', is an interesting shape. The right-handed version is a figure which appears as the symbol of the Isle of Man and also on the flag of Sicily.

timeline

AD 1832	1854	1872
Galois proposes the idea of groups of permutations	Cayley attempts to generalize the concept of a group	Felix Klein begins a programme classifying geometry using group

If we rotate it through 120 degrees or 240 degrees the rotated figure will coincide with itself; if you closed your eyes before rotating it you would see the same triskelion when you opened them again after rotation.

The curious thing about the three-legged figure is that no amount of rotation keeping in the plane will ever convert a right-handed triskelion into a left-handed one. Objects for which the image in the mirror is distinct from the object in front of the mirror are called chiral – they look similar but are not the same. The molecular structure of some chemical compounds may exist in both right-handed and left-handed forms in three dimensions and are examples of chiral objects. This is the case with the compound limosene which in one form tastes like lemons and in the other like oranges. The drug thalidomide in one form is an effective cure of morning sickness in pregnancy but in the other form has tragic consequences.

The Isle of Man triskelion

Measuring symmetry

In the case of our triskelion the basic symmetry operations are the (clockwise) rotations R through 120 degrees and S through 240 degrees. The transformation I is the one that rotates the triangle through 360 degrees or, alternatively, does nothing at all. We can create a table based on the combinations of these rotations, in the same way we might create a multiplication table.

This table is like an ordinary multiplication table with numbers except we are 'multiplying' symbols. According to the most widely used convention, the multiplication $R \circ S$ means first rotate the triskelion clockwise through 240 degrees with S and then by 120 degrees with R, the result being a rotation by 360 degrees, as if you did nothing at all. This can be expressed as $R \circ S = I$, the result found at the junction of the last but one row and the last column of the table.

$\circ$	I	R	S
I	I	R	S
R	R	S	I
S	S	I	R

Cayley table for the symmetry group of the triskelion

The symmetry group of the triskelion is made up of I, R and S and the multiplication table of how to combine them. Because the group contains three elements its size (or 'order') is three. The table is also called a Cayley table (named after the mathematician Arthur Cayley, distant cousin to Sir George Cayley a pioneer of flight).

∘	I	R	S	U	V	W
I	I	R	S	U	V	W
R	R	S	I	V	W	U
S	S	I	R	W	U	V
U	U	W	V	I	S	R
V	V	U	W	R	I	S
W	W	V	U	S	R	I

Caley table for the symmetry group of the tripod

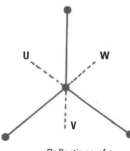

U   W

V

Reflections of a tripod

Like the triskelion, the tripod without feet has rotational symmetry. But it also has mirror symmetry and therefore has a larger symmetry group. We'll call U, V and W the reflections in the three mirror axes.

The larger symmetry group of the tripod, which is of order six, is composed of the six transformations I, R, S, U, V and W and has the multiplication table shown.

An interesting transformation is achieved by combining two reflections in different axes, such as U ∘ W (where the reflection W is applied first and is followed by the reflection U). This is actually a rotation of the tripod through 120 degrees, in symbols U ∘ W = R. Combining the reflections the other way around W ∘ U = S, gives a rotation through 240 degrees. In particular U ∘ W ≠ W ∘ U. This is a major difference between a multiplication table for a group and an ordinary multiplication table with numbers.

A group in which the order of combining the elements is immaterial is called an abelian group, named after the Norwegian mathematician Niels Abel. The symmetry group of the tripod is the smallest group which is not abelian.

Abstract groups The trend in algebra in the 20th century had been towards abstract algebra, in which a group is defined by some basic rules known as axioms. With this viewpoint the symmetry group of the triangle becomes just one example of an abstract system. There are systems in algebra that are more basic than a group and require fewer axioms; other systems that are more complex require more axioms. However the concept of a group is just right and is the most important algebraic system of all. It is remarkable that from so few axioms such a large body of knowledge has emerged. The advantage of the abstract method is that general theorems can be deduced for all groups and applied, if need be, to specific ones.

A feature of group theory is that there may be small groups sitting inside bigger ones. The symmetry group of the triskelion of order three is a subgroup of the symmetry group of the tripod of order six. J.L. Lagrange proved a basic fact about subgroups. Lagrange's theorem states that the order of a subgroup must always divide exactly the order of the group. So we automatically know the symmetry group of the tripod has no subgroups of order four or five.

Classifying groups There has been an extensive programme to classify all the possible finite groups. There is no need to list them all because some groups are built up from basic ones, and it is the basic ones that are needed. The principle of classification is much the same as in chemistry where interest is focused on the basic chemical elements and not the compounds which can be made from them. The symmetry group of the tripod of six elements is a 'compound' being built up from the group of rotations (of order three) and reflections (of order two).

Nearly all basic groups can be classified into known classes. The complete classification, called 'the enormous theorem', was announced by Daniel Gorenstein in 1983 and was arrived at through the accumulated work of 30 years' worth of research and publications by mathematicians. It is an atlas of all known groups. The basic groups fall into one of four main types, yet 26 groups have been found that do not fall into any category. These are known as the sporadic groups.

Axioms for a group

A collection of elements G with 'multiplication' $\circ$ is called a group if

1. There is an element 1 in G so that $1 \circ a = a \circ 1 = a$ for all elements a in the group G (the special element 1 is called the identity element).

2. For each element a in G there is an element $\bar{a}$ in G with $\bar{a} \circ a = a \circ \bar{a} = 1$ (the element $\bar{a}$ is called the inverse element of a).

3. For all elements a, b and c in G it is true that $a \circ (b \circ c) = (a \circ b) \circ c$ (this is called the associative law).

The sporadic groups are mavericks and are typically of large order. Five of the smallest were known to Emile Mathieu in the 1860s but much of the modern activity took place between 1965 and 1975. The smallest sporadic group is of order $7920 = 2^4 \times 3^2 \times 5 \times 11$ but at the upper end are the 'baby monster' and the plain 'monster' which has order $2^{46} \times 3^{20} \times 5^9 \times 7^6 \times 11^2 \times 13^3 \times 17 \times 19 \times 23 \times 29 \times 31 \times 41 \times 47 \times 59 \times 71$ which in decimal speak is around 8×10^{53} or, if you like, 8 with 53 trailing zeros – a very large number indeed. It can be shown that 20 of the 26 sporadic groups are represented as subgroups inside the 'monster' – the six groups that defy all classificatory systems are known as the 'six pariahs'.

Although snappy proofs and shortness are much sought after in mathematics, the proof of the classification of finite groups is something like 10,000 pages of closely argued symbolics. Mathematical progress is not always due to the work of a single outstanding genius.

the condensed idea
Measuring symmetry

39 Matrices

This is the story of 'extraordinary algebra' – a revolution in mathematics which took place in the middle of the 19th century. Mathematicians had played with blocks of numbers for centuries, but the idea of treating blocks as a single number took off 150 years ago with a small group of mathematicians who recognized its potential.

Ordinary algebra is the traditional algebra in which symbols such as a, b, c, x and y represent single numbers. Many people find this difficult to understand, but for mathematicians it was a great step forward. In comparison, 'extraordinary algebra' generated a seismic shift. For sophisticated applications this progress from a one-dimensional algebra to a multiple dimensional algebra would prove incredibly powerful.

Multiple dimensioned numbers In ordinary algebra a might represent a number such as 7, and we would write $a = 7$, but in matrix theory a *matrix* A would be a 'multiple dimensioned number' for example the block

$$A = \begin{pmatrix} 7 & 5 & 0 & 1 \\ 0 & 4 & 3 & 7 \\ 3 & 2 & 0 & 2 \end{pmatrix}$$

This matrix has three rows and four columns (it's a '3 by 4' matrix), but in principle we can have matrices with any number of rows and columns – even a giant '100 by 200' matrix with 100 rows and 200 columns. A critical advantage of matrix algebra is that we can think of vast arrays of numbers, such as a data set in statistics, as a single entity. More than this, we can manipulate these blocks of numbers simply and efficiently. If we want to add or multiply together all the numbers in two data sets, each consisting of 1000 numbers, we don't have to perform 1000 calculations – we just have to perform one (adding or multiplying the two matrices together).

timeline

200BC	AD**1850**	**1858**
Chinese mathematicians use arrays of numbers	J.J. Sylvester introduces the term 'matrix'	Cayley publishes *M* *on the Theory of M.*

A practical example Suppose the matrix A represents the output of the AJAX company in one week. The AJAX company has three factories located in different parts of the country and their output is measured in units (say 1000s of items) of the four products it produces. In our example, the quantities, tallying with matrix A opposite, are:

	product 1	product 2	product 3	product 4
factory 1	7	5	0	1
factory 2	0	4	3	7
factory 3	3	2	0	2

In the next week the production schedule might be different, but it could be written as another matrix B. For example B might be given by

$$B = \begin{pmatrix} 9 & 4 & 1 & 0 \\ 0 & 5 & 1 & 8 \\ 4 & 1 & 1 & 0 \end{pmatrix}$$

What is the total production for both weeks? The matrix theorist says it is the matrix A + B where corresponding numbers are added together,

$$A + B = \begin{pmatrix} 7+9 & 5+4 & 0+1 & 1+0 \\ 0+0 & 4+5 & 3+1 & 7+8 \\ 3+4 & 2+1 & 0+1 & 2+0 \end{pmatrix} = \begin{pmatrix} 16 & 9 & 1 & 1 \\ 0 & 9 & 4 & 15 \\ 7 & 3 & 1 & 2 \end{pmatrix}$$

Easy enough. Sadly, matrix multiplication is less obvious. Returning to the AJAX company, suppose the unit profit of its four products are **3, 9, 8, 2**. We can certainly compute the overall profit for Factory 1 with outputs 7, 5, 0, 1 of its four products. It works out as $7 \times 3 + 5 \times 9 + 0 \times 8 + 1 \times 2 = 68$.

But instead of dealing with just one factory we can just as easily compute the total profits T for *all* the factories

$$T = \begin{pmatrix} 7 & 5 & 0 & 1 \\ 0 & 4 & 3 & 7 \\ 3 & 2 & 0 & 2 \end{pmatrix} \times \begin{pmatrix} 3 \\ 9 \\ 8 \\ 2 \end{pmatrix} = \begin{pmatrix} 7\times3 + 5\times9 + 0\times8 + 1\times2 \\ 0\times3 + 4\times9 + 3\times8 + 7\times2 \\ 3\times3 + 2\times9 + 0\times8 + 2\times2 \end{pmatrix} = \begin{pmatrix} 68 \\ 74 \\ 31 \end{pmatrix}$$

1878

Georg Frobenius proves some of the key results of matrix algebra

1925

Heisenberg uses matrix mechanics in quantum theory

Look carefully and you'll see the *row* by *column* multiplication, an essential feature of matrix multiplication. If in addition to the unit profits we are given the unit *volumes* **7, 4, 1, 5** of each unit of the products, in one fell swoop we can calculate the profits *and* storage requirements for the three factories by the single matrix multiplication:

$$\begin{pmatrix} 7 & 5 & 0 & 1 \\ 0 & 4 & 3 & 7 \\ 3 & 2 & 0 & 2 \end{pmatrix} \times \begin{pmatrix} 3 & 7 \\ 9 & 4 \\ 8 & 1 \\ 2 & 5 \end{pmatrix} = \begin{pmatrix} 68 & 74 \\ 74 & 54 \\ 31 & 39 \end{pmatrix}$$

The total storage is provided by the second column of the resulting matrix, that is 74, 54 and 39. Matrix theory is very powerful. Imagine the situation of a company with hundreds of factories, thousands of products, and different unit profits and storage requirements in different weeks. With matrix algebra the calculations, and our understanding, are fairly immediate, without having to worry about the details which are all taken care of.

Matrix algebra vs ordinary algebra There are many parallels to be drawn between matrix algebra and ordinary algebra. The most celebrated difference occurs in the multiplication of matrices. If we multiply matrix A with matrix B and then try it the other way round:

$$A \times B = \begin{pmatrix} 3 & 5 \\ 2 & 1 \end{pmatrix} \times \begin{pmatrix} 7 & 6 \\ 4 & 8 \end{pmatrix} = \begin{pmatrix} 3 \times 7 + 5 \times 4 & 3 \times 6 + 5 \times 8 \\ 2 \times 7 + 1 \times 4 & 2 \times 6 + 1 \times 8 \end{pmatrix} = \begin{pmatrix} 41 & 58 \\ 18 & 20 \end{pmatrix}$$

$$B \times A = \begin{pmatrix} 7 & 6 \\ 4 & 8 \end{pmatrix} \times \begin{pmatrix} 3 & 5 \\ 2 & 1 \end{pmatrix} = \begin{pmatrix} 7 \times 3 + 6 \times 2 & 7 \times 5 + 6 \times 1 \\ 4 \times 3 + 8 \times 2 & 4 \times 5 + 8 \times 1 \end{pmatrix} = \begin{pmatrix} 33 & 41 \\ 28 & 28 \end{pmatrix}$$

So in matrix algebra we may have A × B and B × A being different, a situation which does not arise in ordinary algebra where the order of multiplying two numbers together makes no difference to the answer.

Another difference occurs with inverses. In ordinary algebra inverses are easy to calculate. If $a = 7$ its inverse is ⅐ because it has the property that ⅐ × 7 = 1. We sometimes write this inverse as $a^{-1} = $ ⅐ and we have $a^{-1} \times a = 1$.

An example in matrix theory is $A = \begin{pmatrix} 1 & 2 \\ 3 & 7 \end{pmatrix}$ and we can verify that $A^{-1} = \begin{pmatrix} 7 & -2 \\ -3 & 1 \end{pmatrix}$

because $A^{-1} \times A = \begin{pmatrix} 7 & -2 \\ -3 & 1 \end{pmatrix} \times \begin{pmatrix} 1 & 2 \\ 3 & 7 \end{pmatrix} = \begin{pmatrix} 1 & 0 \\ 0 & 1 \end{pmatrix}$

where $I = \begin{pmatrix} 1 & 0 \\ 0 & 1 \end{pmatrix}$ is called the identity matrix and is the matrix counterpart
of 1 in ordinary algebra. In ordinary algebra, only 0 does not have an
inverse but in matrix algebra many matrices do not have inverses.

Travel plans Another example of using matrices is in the analysis of a
flight network for airlines. This will involve both airport hubs and smaller
airports. In practice this may involve hundreds of destinations – here we'll
look at a small example: the hubs London (**L**) and Paris (**P**), and smaller
airports Edinburgh (**E**), Bordeaux (**B**), and Toulouse (**T**) and the network
showing possible *direct* flights. To use a computer to analyse such networks,
they are first coded using matrices. If there is a direct flight between airports
a 1 is recorded at the intersection of the row and column labelled by these
airports (like from London to Edinburgh). The 'connectivity' matrix which
describes the network above is A.

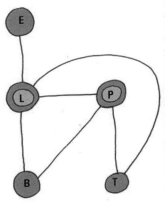

The lower submatrix (marked out by the dotted lines) shows there are no
direct links between the three smaller airports. The matrix product A × A
= A^2 of this matrix with itself can be interpreted as giving the number of
possible journeys between two airports *with exactly one stopover*. So, for
example, there are 3 possible roundtrips to Paris via other cities but no
trips from London to Edinburgh which involve stopovers. The number of
routes which are either direct *or* involve one stopover are the elements of
the matrix $A + A^2$. This is another example of the ability of matrices to
capture the essence of a vast amount of data under the umbrella of a
single calculation.

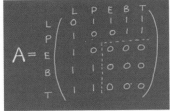

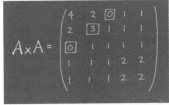

When a small group of mathematicians created the theory of matrices in
the 1850s they did so to solve problems in pure mathematics. From an
applied perspective, matrix theory was very much a 'solution looking for
a problem'. As so often happens 'problems' did arise which needed the
nascent theory. An early application occurred in the 1920s when
Werner Heisenberg investigated 'matrix mechanics', a part of quantum theory.
Another pioneer was Olga Taussky-Todd, who worked for a period in aircraft
design and used matrix algebra. When asked how she discovered the subject
she replied that it was the other way around, matrix theory had found her.
Such is the mathematical game.

the condensed idea
Combining blocks of numbers

40 Codes

What does Julius Caesar have in common with the transmission of modern digital signals? The short answer is codes and coding. To send digital signals to a computer or a digital television set, the coding of pictures and speech into a stream of zeros and ones – a binary code – is essential for it is the only language these devices understand. Caesar used codes to communicate with his generals and kept his messages secret by changing around the letters of his message according to a key which only he and they knew.

Accuracy was essential for Caesar and it is also required for the effective transmission of digital signals. Caesar also wanted to keep his codes to himself as do the cable and satellite broadcasting television companies who only want paying subscribers to be able make sense of their signals.

Let's look at accuracy first. Human error or 'noise along the line' can always occur, and must be dealt with. Mathematical thinking allows us to construct coding systems that automatically detect errors and even make corrections.

Error detection and correction One of the first binary coding systems was the Morse code which makes use of two symbols, dots • and dashes –. The American inventor Samuel F.B. Morse sent the first intercity message using his code from Washington to Baltimore in 1844. It was a code designed for the electric telegraph of the mid-19th century with little thought to an efficient design. In Morse code, the letter A is coded as • –, B as – •••, C as – • – • and other letters as different sequences of dots and dashes. A telegraph operator sending 'CAB' would send the string – • – • / • – / – •••. Whatever its merits, Morse code is not very good at error detection let alone correction. If the Morse code operator wished to send 'CAB', but mistyped a dot for a dash in C, forgot the dash in A and noise on the wire substituted a dash for a dot in B, the receiver getting •• – • / • / – – ••, would see nothing wrong and interpret it as 'FEZ'.

timeline

55BC
Julius Caesar invades Britain and uses codes to communicate with his generals

C.AD**1750**
Euler's theorem lays the foundations for public key cryptography

At a more primitive level we could look at a coding system consisting of just 0 and 1 where 0 represents one word and 1 another. Suppose an army commander has to transmit a message to his troops which is either 'invade' or 'do not invade'. The 'invade' instruction is coded by '1' and the 'do not invade' instruction by '0'. If a 1 or a 0 was incorrectly transmitted the receiver would never know – and the wrong instruction would be given, with disastrous consequences.

We can improve matters by using code words of length two. If this time we code the 'invade' instruction by 11 and the 'do not invade' by 00, this is better. An error in one digit would result in 01 or 10 being received. As only 11 or 00 are legitimate code words, the receiver would certainly know that an error had been made. The advantage of this system is that an error would be detectable, but we still would not know how to correct it. If 01 were received, how would we know whether 00 or 11 should have been sent?

The way to a better system is to combine design with longer code words. If we code the 'invade' instruction by 111 and the 'do not invade' by 000 an error in one digit could certainly be detected, as before. If we knew that at most one error could be made (a reasonable assumption since the chance of two errors in one code word is small), the correction could actually be made by the receiver. For example, if 110 were received then the correct message would have been 111. With our rules, it could not be 000 since this code word is two errors away from 110. In this system there are only two code words 000 and 111 but they are far enough apart to make error detection and correction possible.

The same principle is used when word processing is in autocorrect mode. If we type 'animul' the word processor detects the error and corrects it by taking the nearest word, 'animal'. The English language is not fully correcting though because if we type 'lomp' there is no unique nearest word; the words, lamp, limp, lump, pomp and romp are all equidistant in terms of single errors from lomp.

A modern binary code consists of code words that are blocks consisting of zeros and ones. By choosing the legitimate code words far enough apart, both detection and correction are possible. The code words of Morse code are too close together but the modern code systems used to transmit data from satellites can always be set in autocorrect mode. Long code words with high

844	1920s	1950	1970s
orse transmits the first essage using his code	The Enigma machine is developed	Richard Hamming publishes a key paper on error-detecting and error-correcting codes	Public key cryptography is developed

performance in terms of error correction take longer to transmit so there is a tradeoff between length and speed of transmission. Voyages into deep space by NASA have used codes that are three-error correcting and these have proved satisfactory in combating noise on the line.

Making messages secret Julius Caesar kept his messages secret by changing around the letters of his message according to a key that only he and his generals knew. If the key fell into the wrong hands his messages could be deciphered by his enemies. In medieval times, Mary Queen of Scots sent secret messages in code from her prison cell. Mary had in mind the overthrow of her cousin, Queen Elizabeth, but her coded messages were intercepted. More sophisticated than the Roman method of rotating all letters by a key, her codes were based on substitutions but ones whose key could be uncovered by analysing the frequency of letters and symbols used. During the Second World War the German Enigma code was cracked by the discovery of its key. In this case it was a formidable challenge but the code was always vulnerable because the key was transmitted as part of the message.

A startling development in encryption of messages was discovered in the 1970s. Running counter to everything that had been previously believed, it said that the secret key could be broadcast to all and yet the message could remain entirely safe. This is called public key cryptography. The method depends on a 200 year old theorem in a branch of mathematics glorified for being the most useless of all.

Public key encryption Mr John Sender, a secret agent known in the spying fraternity as 'J', has just arrived in town and wants to send his minder Dr Rodney Receiver a secret message to announce his arrival. What he does next is rather curious. He goes to the public library takes a town directory off the shelf and looks up Dr R. Receiver. In the directory he finds two numbers alongside Receiver's name – a long one, which is 247, and a short one 5. This information is available to all and sundry, and it is all the information John Sender requires to encrypt his message, which for simplicity is his calling card, J. This letter is number 74 in a list of words, again publicly available.

Sender encrypts 74 by calculating 74^5 (modulo 247), that is, he wants to know the remainder on dividing 74^5 by 247. Working out 74^5 is just about possible on a handheld calculator, but it has to be done exactly:

$$74^5 = 74 \times 74 \times 74 \times 74 \times 74 = 2,219,006,624$$

and

$$2,219,006,624 = 8,983,832 \times 247 + 120$$

so dividing his huge number by 247 he gets the remainder 120. Sender's encrypted message is 120 and he transmits this to Receiver. Because the numbers 247 and 5 were publicly available anyone could encrypt a message. But not everyone could decrypt it. Dr R. Receiver has more information up his sleeve. He made up his personal number 247 by multiplying together two prime numbers. In this case he obtained the number 247 by multiplying $p = 13$ and $q = 19$, but only he knows this.

This is where the ancient theorem due to Leonhard Euler is taken out and dusted down. Dr R. Receiver uses the knowledge of $p = 13$ and $q = 19$ to find a value of a where $5 \times a \equiv 1$ modulo $(p-1)(q-1)$ where the symbol $\equiv$ means equals in modular arithmetic. What is a so that dividing $5 \times a$ by $12 \times 18 = 216$ leaves remainder 1? Skipping the actual calculation he finds $a = 173$.

Because he is the only one who knows the prime numbers p and q, Dr Receiver is the only one who can calculate the number 173. With it he works out the remainder when he divides the huge number 120^{173} by 247. This is outside the capacity of a hand held calculator but is easily found by using a computer. The answer is 74, as Euler knew two hundred years ago. With this information, Receiver looks up word 74 and sees that J is back in town.

You might say, surely a hacker could discover the fact that $247 = 13 \times 19$ and the code could be cracked. You would be correct. But the encryption and decryption principle is the same if Dr Receiver had used another number instead of 247. He could choose two very big prime numbers and multiply them together to get a much larger number than 247.

Finding the two prime factors of a very large number is virtually impossible – what are the factors of 24,812,789,922,307 for example? But numbers much larger than this could also be chosen. The public key system is secure and if the might of supercomputers joined together are successful in factoring an encryption number, all Dr Receiver has to do is increase its size still further. In the end it is considerably easier for Dr Receiver to 'mix boxes of black sand and white sand together' than for any hacker to unmix them.

the condensed idea
Keeping messages secret

41 Advanced counting

The branch of mathematics called combinatorics is sometimes known as advanced counting. It is not about adding up a column of figures in your head. 'How many?' is a question, but so is 'how can objects be combined?' Problems are often simply stated, unaccompanied by the weighty superstructure of mathematical theory – you don't have to know a lot of preliminary work before you can roll up your sleeves. This makes combinatorial problems attractive. But they should carry a health warning: addiction is possible and they can certainly cause lack of sleep.

A tale from St Ives Children can start combinatorics at a tender age. One traditional nursery rhyme poses a combinatorial question:

> As I was going to St Ives,
> I met a man with seven wives;
> Every wife had seven sacks,
> Every sack had seven cats,
> Every cat had seven kits.
> Kits, cats, sacks and wives,
> How many were going to St Ives?

The last line contains the trick question. An implicit assumption is that the narrator is the only person on his way 'to' St Ives, so the answer is 'one'. Some people exclude the narrator and for them the answer would be 'none'.

The charm of the poem lies in its ambiguity and the various questions it can generate. We could ask: how many were coming *from* St Ives? Again interpretation is important. Can we be sure the man with his seven wives were all travelling *away*

timeline

c.1800 BC	c.AD 1100
The Rhind papyrus is written in Egypt	Bhaskara deals with permutations and combinations

from St Ives? Were the wives accompanying the man when he was met, or were they somewhere else? The first requirement of a combinatorial problem is that assumptions be agreed beforehand.

We'll assume the entourage was coming along the single road away from the Cornish seaside town and that the 'kits, cats, sacks and wives' were all present. How many were there coming from St Ives? The following table gives us a solution.

man	1	1
wives	7	7
sacks	7×7	49
cats	$7 \times 7 \times 7$	343
kits	$7 \times 7 \times 7 \times 7$	2401
Total		**2801**

In 1858 Alexander Rhind a Scottish antiquarian visiting Luxor came across a 5 metre long papyrus filled with Egyptian mathematics from the period 1800 BC. He bought it. A few years later it was acquired by the British Museum and its hieroglyphics translated. Problem 79 of the Rhind Papyrus is a problem of houses, cats, mice and wheat very similar to the kits, cats, sacks and wives of St Ives. Both involve powers of 7 and the same kind of analysis. Combinatorics, it seems, has a long history.

Factorial numbers The problem of queues introduces us to the first weapon in the combinatorial armoury – the *factorial* number. Suppose **A**lan, **B**rian, **C**harlotte, **D**avid, and **E**llie form themselves into a queue

E   C   A   B   D

with **E**llie at the head of the queue followed by **C**harlotte, **A**lan and **B**rian with **D**avid at the end. By swapping the people around other queues are formed; how many different queues are possible?

The art of counting in this problem depends on *choice*. There are 5 choices for who we place as the first person in the queue, and once this person has been chosen, there are 4 choices for the second person, and so on. When we come to the last position there is no choice at all as it can only be filled by the person left over. There are therefore $5 \times 4 \times 3 \times 2 \times 1 = 120$ possible queues. If we started with 6 people, the number of different queues would be $6 \times 5 \times 4 \times 3 \times 2 \times 1 = 720$ and for 7 people there would be $7 \times 6 \times 5 \times 4 \times 3 \times 2 \times 1 = 5040$ possible queues.

1971

Ray-Chaudhuri and Wilson
prove the existence of general
Kirkman's systems

A number obtained by multiplying successive whole numbers is called a factorial number. These occur so often in mathematics that they are written using the notation 5! (read '5 factorial') instead of $5 \times 4 \times 3 \times 2 \times 1$. Let's take a look at the first few factorials (we'll define 0! to equal 1). Straightaway, we see that quite 'small' configurations give rise to 'large' factorial numbers. The number n may be small but $n!$ can be huge.

number	factorial
0	1
1	1
2	2
3	6
4	24
5	120
6	720
7	5040
8	40,320
9	362,880

If we're still interested in forming queues of 5 people, but can now draw on a pool of 8 people **A**, **B**, **C**, **D**, **E**, **F**, **G**, and **H**, the analysis is almost the same. There are 8 choices for the front person in the queue, 7 for the second and so on. But this time there are 4 choices for the last slot. The number of possible queues is

$$8 \times 7 \times 6 \times 5 \times 4 = 6720$$

This can be written with the notation for factorial numbers, because

$$8 \times 7 \times 6 \times 5 \times 4 = 8 \times 7 \times 6 \times 5 \times 4 \times \frac{3 \times 2 \times 1}{3 \times 2 \times 1} = \frac{8!}{3!}$$

Combinations In a queue the *order* matters. The two queues

$$\text{C} \quad \text{E} \quad \text{B} \quad \text{A} \quad \text{D} \qquad\qquad \text{D} \quad \text{A} \quad \text{C} \quad \text{E} \quad \text{B}$$

are made from the same letters but are different queues. We already know there are 5! queues that can be made with these letters. If we're interested in counting the ways of selecting 5 people from 8 *immaterial of order* we must divide $8 \times 7 \times 6 \times 5 \times 4 = 6720$ by 5!. The number of ways of selecting 5 people from 8 is therefore

$$\frac{8 \times 7 \times 6 \times 5 \times 4}{5 \times 4 \times 3 \times 2 \times 1} = 56$$

This number, using C for combination, is written $^8C_5$ and is

$$^8C_5 = \frac{8!}{3!5!} = 56$$

In the UK National Lottery the rules require a selection of 6 numbers from a possible 49 – how many possibilities are there?

$$^{49}C_6 = \frac{49!}{43!6!} = \frac{49 \times 48 \times 47 \times 46 \times 45 \times 44}{6 \times 5 \times 4 \times 3 \times 2 \times 1} = 13,983,816$$

Only one combination wins so there is approximately 1 chance in 14 million of picking the jackpot.

Kirkman's problem Combinatorics is a wide field and, though old, it has rapidly developed over the past 40 years, due to its relevance to computer science. Problems involving graph theory, Latin squares and the like can be thought of as part of modern combinatorics.

The essence of combinatorics is captured by a master of the subject, Rev. Thomas Kirkman, working at a time when combinatorics was mostly linked to recreational mathematics. He made many original contributions to discrete geometry, group theory and combinatorics but never had a university appointment. The conundrum which reinforced his reputation as a no-nonsense mathematician was the one for which he will always be known. In 1850 Kirkman introduced the '15 schoolgirls problem', in which schoolgirls walk to church in 5 rows of 3 on each day of the week. If you are bored with Sudoku you might try to solve it. We need to organize a daily schedule so that no two walk together more than once. Using lower case and upper case deliberately, the girls are: abigail, beatrice, constance, dorothy, emma, frances, grace, Agnes, Bernice, Charlotte, Danielle, Edith, Florence, Gwendolyn and Victoria, labelled a, b, c, d, e, f, g, A, B, C, D, E, F, G and V, respectively.

There are actually seven distinct solutions to Kirkman's problem, and the one we'll give is 'cyclic' – it is generated by 'going around'. This is where the labelling of the schoolgirls comes into its own.

Monday			Tuesday			Wednesday			Thursday			Friday			Saturday			Sunday		
a	A	V	b	B	V	c	C	V	d	D	V	e	E	V	f	F	V	g	G	V
b	E	D	c	F	E	d	G	F	e	A	G	f	B	A	g	C	B	a	D	C
c	B	G	d	C	A	e	D	B	f	E	C	g	F	D	a	G	E	b	A	F
d	f	g	e	g	a	f	a	b	g	b	c	a	c	d	b	d	e	c	e	f
e	F	C	f	G	D	g	A	E	a	B	F	b	C	G	c	D	A	d	E	B

It is called cyclic since on each subsequent day the walking schedule is changed from **a** to **b**, **b** to **c**, down to **g** to **a**. The same applies to the upper-case girls **A** to **B**, **B** to **C**, and so on, but Victoria remains unmoved.

The underlying reason for the choice of notation is that the rows correspond to lines in the Fano geometry (see page 115). Kirkman's problem isn't only a parlour game but one that's part of mainstream mathematics.

the condensed idea
How many combinations?

42 Magic squares

'A Mathematician', wrote G.H. Hardy, 'like a painter or a poet, is a maker of patterns.' Magic squares have very curious patterns even by mathematical standards. They lie on the border between heavily symbolled mathematics and the fascinating patterns loved by puzzlesmiths.

A magic square is a square grid in which distinct whole numbers are written into each cell of the grid in such a way that each horizontal row and each vertical column, *and* each diagonal add up to the same number.

Squares with just one row and one column are technically magic squares but are very boring so we'll forget them. There is no such thing as a magic square with two rows and two columns. If there were it would have the form shown. Since the row additions and the column additions should be equal, then $a + b = a + c$. This means $b = c$, *contradicting* the fact that all the entries must be distinct.

The Lo Shu square As 2×2 squares do not exist, we'll look at 3×3 arrays and attempt to construct one with a grid. We'll start with a *normal* magic square, one where the grid is filled out with the consecutive numbers 1, 2, 3, 4, 5, 6, 7, 8 and 9.

For such a small square it is possible to construct a 3×3 magic square by the 'trial and test' method, but we can first make some deductions to help us along. If we add up *all* the numbers in the grid we have

$$1 + 2 + 3 + 4 + 5 + 6 + 7 + 8 + 9 = 45$$

and this total would have to be the same as adding the totals of 3 rows. This shows that each row (and column and diagonal) must add up to 15. Now let's look at the middle cell – we'll call this *c*. Two diagonals involve *c* as does the middle row and the middle column. If we add the numbers in these four lines together we get 15 +15 +15 +15 = 60 and this must equal *all* the numbers added together plus 3 extra lots of *c*. From the equation $3c + 45 = 60$, we see that *c must* be 5. Other facts can also be learned such as not being able to place a 1 in a corner cell. Some clues gathered, we are in a good position to use the trial and test method. Try it!

Of course we'd like a totally systematic *method* for constructing magic squares. One was found by Simon de la Loubère, the French ambassador to the King of Siam in the late 17th century. Loubère took an interest in Chinese mathematics and wrote down a method for constructing magic squares that have an odd number of rows and columns. This method starts by placing a 1 in the middle of the first row and 'going up and across and rotating if necessary' to place the 2 and subsequent numbers. If blocked the next number beneath the current number is used.

8	1	6
3	5	7
4	9	2

A solution for the 3×3 square by the Siamese method

Remarkably this normal magic square is essentially the only one with 3 rows and 3 columns. Every other 3×3 magic square can be obtained from this one by rotating numbers about the middle and/or reflecting numbers of the square in the middle column or middle row. It is called the 'Lo Shu' square and was known in China around 3000 BC. Legend says that it was first seen on the back of a turtle emerging from the Lo river. The local people took this as a sign from the gods that they would not be freed of pestilence unless they increased their offerings.

If there is one 3×3 magic square, how many distinct 4×4 magic squares are there? The staggering answer is that there are 880 different ones (and be prepared, there are 2,202,441,792 magic squares of order 5). We don't know how many squares there are for general values of *n*.

Dürer and Franklin's squares
The Lo Shu magic square is well known for its age and uniqueness but one 4×4 magic square has become iconic for its association with a famous artist. It also has many more properties than some run of the mill magic squares that make up the 880 different versions. This is the 4×4 square in Albrecht Dürer's engraving of *Melancholia*, which he made in the year 1514.

1693	**1770**	**1986**
Bernard Frénicle de Bessy lists all the 880 possible 4×4 magic squares	Euler produces a squared square	Sallows creates his letter-based square

In Dürer's square all the rows add up to 34, as do the columns, the diagonals, and the 2×2 small squares which make up the complete 4×4 square. Dürer even managed to 'sign' his masterpiece with the date of its completion in the middle of the lowest row.

The American scientist and diplomat Benjamin Franklin saw that constructing magic squares was a useful tool for sharpening the mind. He was adept at this, and to this day mathematicians have little idea how he did it; large magic squares cannot be constructed by serendipity. Franklin confessed that in his youth he had wasted much time over them despite not being taken with the 'Arithmetick' as a boy. Here's one he discovered in his youth.

52	61	4	13	20	29	36	45
14	3	62	51	46	35	30	19
53	60	5	12	21	28	37	44
11	6	59	54	43	38	27	22
55	58	7	10	23	26	39	42
9	8	57	56	41	40	25	24
50	63	2	15	18	31	34	47
16	1	64	49	48	33	32	17

In this normal magic square there are all kinds of symmetries. All the rows, columns and diagonals add up to 260, as do the 'bent rows', one of which we've highlighted. There are many other things to discover – like the sum of the central 2×2 square plus the four corner boxes, which also adds up to 260. Look closely and you'll find an interesting result for every 2×2 square.

Squared squares Some magic squares can have cells occupied by different squared numbers. The problem of constructing these was posed by the French mathematician Edouard Lucas in 1876. To date no 3×3 square of squares has been found, although one has come close.

127^2	46^2	58^2
2^2	113^2	94^2
74^2	82^2	97^2

All rows and columns and *one* diagonal of this square add up to the magic sum 21,609 but the other diagonal fails since $127^2 + 113^2 + 97^2 = 38,307$. If you're tempted to find one yourself you should take note of a proven result: the centre cell value must be greater than 2.5×10^{25} so there's little point in looking for a square with small numbers! This is serious mathematics which has a connection with elliptic curves, the topic used to prove Fermat's Last Theorem. It has been proved there are no 3×3 magic squares whose entries are cubes or fourth powers.

The search for squared squares has, however, been successful for larger squares. Magic 4×4 and 5×5 squared squares do exist. In 1770 Euler produced an example without showing his method of construction. Whole families have since been found linked to the study of the algebra of quaternions, the four-dimensional imaginary numbers.

Exotic magic squares Large magic squares may have spectacular properties. A 32×32 array has been produced by magic square expert William Benson in which the numbers, their squares, and their cubes all form magic squares. In 2001 a 1024×1024 square was produced in which all powers of elements up to the fifth power make magic squares. There are many results like these.

We can create a whole variety of other magic squares if the requirements are relaxed. The normal magic squares are in the mainstream. Removing the condition that the sum of the diagonal elements must equal the sums of the rows, and of the columns, ushers in a plethora of specialized results. We can search for squares whose elements consist only of prime numbers, or we may consider shapes other than squares which have 'magic properties'. By going into higher dimensions we are led to consider magic cubes and hypercubes.

But the prize for the most remarkable magic square of all, certainly for curiosity value, must go to a humble 3×3 square produced by the Dutch electronic engineer and wordsmith Lee Sallows:

5	22	18
28	15	2
12	8	25

What is so remarkable about this? First write out the numbers in words:

five	twenty-two	eighteen
twenty-eight	fifteen	two
twelve	eight	twenty-five

Then count the number of *letters* making up each word to obtain:

4	9	8
11	7	3
6	5	10

Remarkably, this is a magic square consisting of the consecutive numbers 3, 4, 5, up to 11. We also find that the number of letters of the magic sums of both 3×3 squares (21 and 45) is 9 and fittingly 3 × 3 = 9.

the condensed idea
Mathematical wizardry

43 Latin squares

For a few years the world has been Sudoku mad. Across the land, pens and pencils are chewed waiting for the right inspiration for the number to put in that box. Is it 4 or is it 5? Maybe it's 9. Commuters emerge from their trains in the mornings having expended more mental effort than they will for the rest of the day. In the evening the dinner burns in the oven. Is it 5, 4, or maybe 7? All these people are playing with Latin squares – they are being mathematicians.

	4		8		3			
		7						3
		9	7			2	6	
3				1		7		9
		6	9	8				
1		5		2				6
	2	3			6	5		
6						1		
		5		2		8		

Sudoku unlocked In Sudoku we are given a 9×9 grid with some numbers filled in. The object is to fill in the rest using the given numbers as clues. Each row and each column should contain exactly one of the digits 1, 2, 3, …, 9, as do the small constituent 3×3 squares.

It is believed that Sudoku (meaning 'single digits') was invented in the late 1970s. It gained popularity in Japan in the 1980s before sweeping to mass popularity by 2005. The appeal of the puzzle is that, unlike crosswords, you don't have to be widely read to attempt them but, like crosswords, they can be compelling. Addicts of both forms of self-torture have much in common.

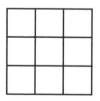

3×3 Latin squares A square array containing exactly one symbol in each row and each column is called a Latin square. The number of symbols equals the size of the square and is called its 'order'. Can we fill out a blank 3×3 grid so that each row and column contains exactly one of the symbols *a*, *b* and *c*? If we can, this would be a Latin square of order 3.

In introducing the concept of a Latin square, Leonhard Euler called it a 'new kind of magic square'. Unlike magic squares, however, Latin squares are not

timeline

AD **1779**

Euler explores the theory of Latin squares

1900

Tarry shows there are no orthogonal Latin squares of order 6

concerned with arithmetic and the symbols do not have to be numbers. The reason for the name is simply that the symbols used to form them are taken from the Latin alphabet, while Euler used Greek with other squares.

a	b	c
b	c	a
c	a	b

A 3×3 Latin square can be easily written down.

If we think of *a*, *b* and *c* as the days of the week Monday, Wednesday and Friday, the square could be used to schedule meetings between two teams of people. Team One is made up of **L**arry, **M**ary and **N**ancy and Team Two of **R**oss, **S**ophie and **T**om.

	R	S	T
L	a	b	c
M	b	c	a
N	c	a	b

For example, **M**ary from Team One, has a meeting with **T**om from Team Two on Monday (the intersection of the **M** row with the **T** column is *a* = Monday). The Latin square arrangement ensures a meeting takes place between each pair of team members and there is no clash of dates.

This is not the only possible 3×3 Latin square. If we interpret A, B and C as topics discussed at the meetings between Team One and Team Two, we can produce a Latin square which ensures each person discusses a different topic with a member of the other team.

	R	S	T
L	A	B	C
M	C	A	B
N	B	C	A

So **M**ary from Team One discusses topic C with **R**oss, topic A with **S**ophie and topic B with **T**om.

But *when* should the discussions take place, between *who*, and on *what* topic? What would be the schedule for this complex organization? Fortunately the two Latin squares can be combined symbol by symbol to produce a composite Latin Square in which each of the possible nine pairs of days and topics occurs in exactly one position.

sher suggests using Latin quares to design statistical xperiments

Euler's conjecture about the non-existence of certain pairs of Latin squares is disproved by Bose, Parker and Shrikhande

Sudoku-like games are invented in New York

	R	S	T
L	*a,A*	*b,B*	*c,C*
M	*b,C*	*c,A*	*a,B*
N	*c,B*	*a,C*	*b,A*

Another interpretation for the square is the historical 'nine officers problem' in which nine officers belonging to three regiments *a*, *b* and *c* and of three ranks A, B and C are placed on the parade ground so that each row and column contains an officer of each regiment and rank. Latin squares which combine in this way are called 'orthogonal'. The 3×3 case is straightforward but finding pairs of orthogonal Latin squares for some larger ones is far from easy. This is something Euler discovered.

In the case of a 4×4 Latin Square, a '16 officers problem' would be to arrange the 16 court cards in a pack of cards in a square in such a way that there is one rank (Ace, King, Queen or Jack) and one suit (spades, clubs, hearts or diamonds) in each row and column. In 1782 Euler posed the same problem for '36 officers'. In essence he was looking for two orthogonal squares of order 6. He couldn't find them and conjectured there were *no* pairs of orthogonal Latin squares of orders 6, 10, 14, 18, 22 … Could this be proved?

Along came Gaston Tarry, an amateur mathematician who worked as a civil servant in Algeria. He scrutinized examples and by 1900 had verified Euler's conjecture in one case: there is no pair of orthogonal Latin squares of order 6. Mathematicians naturally assumed Euler was correct in the other cases 10, 14, 18, 22 …

In 1960, the combined efforts of three mathematicians stunned the mathematical world by proving Euler wrong in *all* the other cases. Raj Bose, Ernest Parker and Sharadchandra Shrikhande proved there were indeed pairs of orthogonal Latin squares of orders 10, 14, 18, 22, … The *only* case where Latin squares do not exist (apart from trivial ones of orders 1 and 2) is order 6.

We've seen that there are two mutually orthogonal order 3 Latin squares. For order 4 we can produce three squares which are mutually orthogonal to each other. It can be shown that there are never more than $n - 1$ mutually orthogonal Latin squares of order n, so for $n = 10$, for example, there cannot be more than nine mutually orthogonal squares. But finding them is a different story. To date, no one has been able to even produce three Latin squares of order 10 that are mutually orthogonal to each other.

Are Latin squares useful? R.A. Fisher, an eminent statistician, saw the practical use of Latin squares. He used them to revolutionize agricultural methods during his time at Rothamsted Research Station in Hertfordshire, UK.

Fisher's objective was to investigate the effectiveness of fertilizers on crop yield. Ideally we would want to plant crops in identical soil conditions so that soil quality wasn't an unwanted factor influencing crop yield. We could then apply the different fertilizers safe in the knowledge that the 'nuisance' of soil quality was eliminated. The only way of ensuring identical soil conditions would be to use the *same* soil – but it is impractical to keep digging up and replanting crops. Even if this were possible different weather conditions could become a new nuisance.

A way round this is to use Latin squares. Let's look at testing four treatments. If we mark out a square field into 16 plots we can envisage the Latin square as a description of the field where the soil quality varies 'vertically' and 'horizontally'.

The four fertilizers are then applied at random in the scheme labelled a, b, c and d, so exactly one is applied in each row and column in an attempt to remove the variation of soil quality. If we suspect another factor might influence crop yield, we could deal with this too. Suppose we think that the time of day of apllying treatment is a factor. Label four time zones during the day as A, B, C and D and use orthogonal Latin squares as the design for a scheme to gather data. This ensures each treatment and time zone is applied in one of the plots. The design for the experiment would be:

a, time A	b, time B	c, time C	d, time D
b, time C	a, time D	d, time A	c, time B
c, time D	d, time C	a, time B	b, time A
d, time B	c, time A	b, time D	a, time C

Other factors can be screened out by going on to create even more elaborate Latin square designs. Euler could not have dreamt of the solution to his officers' problem being applied to agricultural experiments.

the condensed idea
Sudoku revealed

44 Money mathematics

Norman is a super salesperson when it comes to bikes. He also sees it as his duty to get everyone on a bike, so he is delighted when a customer comes into his shop and without any hesitation buys a bike for £99. The customer pays for it with a cheque for £150, and as the banks are closed, Norman asks his neighbour to cash it. He returns, gives his customer the change of £51 who then rides off at speed. Calamity follows. The cheque bounces, the neighbour demands his money back, and Norman has to go to a friend to borrow the money. The bike originally cost him £79, but how much did Norman lose altogether?

The concept of this little conundrum was proposed by the great puzzlesmith Henry Dudeney. It is money mathematics of a sort, but more accurately a puzzle connected with money. It shows how money depends on time and that inflation is alive and well. Writing in the 1920s, Dudeney's bike actually cost the customer £15. A way to combat inflation is through the interest on money. This is the stuff of serious mathematics and the modern financial market place.

Compound interest There are two sorts of interest, known as simple and compound. Let's turn our mathematical spotlight onto two brothers, Compound Charlie and Simple Simon. Their father gives them each £1000, which they both place in a bank. Compound Charlie always chooses an account that applies compound interest but Simple Simon is more traditional and prefers accounts that use simple interest. Historically, compound interest was identified with usury and frowned upon. Nowadays compound interest is a fact of life, central to modern monetary systems. Compound interest is interest

timeline

3000BC

The Babylonians use a sexagesimal
number system for financial dealings

AD**1494**

Luca Pacioli publishes fi▪
tables and an account of
double-entry bookkeepir

compounded on interest, and that is why Charlie likes it. Simple interest does not have this feature and is calculated on a set amount known as the 'principal'. Simon can understand it easily, as the principal earns the same amount of interest each year.

When talking about mathematics, it is always good to have Albert Einstein on side – but the widespread claim that he said compound interest is the greatest discovery of all time is too far-fetched. That the formula for compound interest has a greater immediacy than his $E = mc^2$ is undeniable. If you save money, borrow money, use a credit card, take out a mortgage or buy an annuity, the compound interest formula is in the background working for (or against) you. What do the symbols stand for? The term P stands for principal (the money you save or borrow), i is the percentage interest rate divided by 100 and n is the number of time periods.

$$A = P \times (1 + i)^n$$

Compound interest formula

Charlie places his £1000 in an account paying 7% interest annually. How much will accrue in three years? Here $P = 1000$, $i = 0.07$ and $n = 3$. The symbol A represents the accrued amount and by the compound interest formula $A = £1225.04$.

Simon's account pays the same interest rate, 7%, as simple interest. How do his earnings compare after three years? For the first year he would gain £70 in interest and this would be the same in the second and third years. He would therefore have 3 × £70 interest giving a total accrued amount of £1210. Charlie's investment was the better business decision.

Sums of money that grow by compounding can increase very rapidly. This is fine if you are saving but not so good if you are borrowing. A key component of compound interest is the period at which the compounding takes place. Charlie has heard of a scheme which pays 1% per week, a penny in every pound. How much would he stand to gain with this scheme?

Simon thinks he knows the answer: he suggests we multiply the interest rate 1% by 52 (the number of weeks in the year) to obtain an annual percentage rate of 52%. This means an interest of £520 making a total of £1520 in the account. Charlie reminds him, however, of the magic of compound interest

1718	**1756**	**1848**
Abraham de Moivre investigates mortality statistics and the foundation of the theory of annuities	James Dodson publishes *First Lectures on Insurances*	The Institute of Actuaries is founded in London

and the compound interest formula. With $P = £1000$, $i = 0.01$ and $n = 52$, Charlie calculates the accrual to be $£1000 \times (1.01)^{52}$. Using his calculator he finds this is £1677.69, much more than the result of Simple Simon's sum. Charlie's equivalent annual percentage rate is 67.769% and is much greater than Simon's calculation of 52%.

Simon is impressed but his money is already in the bank under the simple interest regime. He wonders how long it will take him to double his original £1,000? Each year he gets £70 interest so all he has to do is divide 1000 by 70. This gives 14.29 so that he can be sure in 15 years he will have more than £2000 in the bank. It is a long time to wait. To show the superiority of compound interest Charlie starts to calculate his own doubling period. This is a little more complicated but a friend tells him about the rule of 72.

The rule of 72 For a given percentage rate, the rule of 72 is a rule of thumb for estimating the number of periods required for money to double. Though Charlie is interested in years the rule of 72 applies to days or month as well. To find the doubling period all Charlie has to do is to divide 72 by the interest rate. The calculation is $72/7 = 10.3$ so Charlie can report to his brother that his investment will double in 11 years, much quicker than Simon's 15. The rule is an approximation but it is useful where quick decisions have to be made.

Present value Compound Charlie's father is so impressed by his son's good sense that he takes him aside and says 'I propose to give you £100,000'. Charlie is very excited. Then his father adds the condition that he will only give him the £100,000 when he is 45 and that won't be for another ten years. Charlie is not so happy.

Charlie wants to spend the money now but obviously he cannot. He goes to his bank and promises them the £100,000 in ten years time. The bank responds that time is money and £100,000 in ten years time is not the same as £100,000 now. The bank has to estimate the size of investment now that would realize £100,000 in ten years. This will be the amount they will loan to Charlie. The bank believes that a growth rate of 12% would give them a healthy profit. What would be the amount now that would grow to £100,000 in ten years, at 12% interest? The compound interest formula can be used for this problem as well. This time we are given $A = £100,000$ and have to calculate P, the present value of A. With $n = 10$ and $i = 0.12$, the bank will be prepared to advance Charlie the amount $100,000/1.12^{10} = £32,197.32$. Charlie is quite shocked by this small figure, but he will still be able to buy that new Porsche.

How can regular payments be handled? Now that Charlie's father has promised to give £100,000 to his son in ten years time, he has to save the money up. This he plans to do with a stream of equal saving account payments made at the end of each year for ten years. By the end of this period he will then be able to hand over the money to Charlie on the day he has promised, and Charlie can hand the money to the bank to pay off the loan.

Charlie's father manages to find an account that allows him to do this, an account that pays an annual interest rate of 8% for the whole ten year term. He gives Charlie the task of working out the annual payments. With the compound interest formula Charlie was concerned with one payment (the original principal) but now he is concerned with ten payments made at different times. If regular payments R are made at the end of each year in an environment where the interest rate is i, the amount saved after n years can be calculated by the regular payments formula.

$$S = R \times \frac{((1+i)^n - 1)}{i}$$

Regular payments formula

Charlie knows that S = £100,000, n = 10 and i = 0.08 and calculates that R = £6902.95.

Now that Charlie has his brand new Porsche, courtesy of the bank, he needs a garage to put it in. He decides to take out a mortgage for £300,000 to buy a house, a sum of money he will pay back in a stream of equal annual payments over 25 years. He recognizes this as a problem in which the £300,000 is the present value of a stream of payments to be made and he calculates his annual payments with ease. His father is impressed and makes further use of Charlie's prowess. He has just been given a retirement lump sum of £150,000 and wants to purchase an annuity. 'That's OK,' says Charlie, 'we can use the same formula, as the mathematics is the same. Instead of the mortgage company advancing me money that I repay in regular instalments, you are giving them the money and they are making the regular payments to you.'

By the way, the answer to Henry Dudeney's brainteaser is £130, made up of the £51 Norman gave the customer and the £79 he paid for the bike.

the condensed idea
Compound interest
works best

45 The diet problem

Tanya Smith takes her athletics very seriously. She goes to the gym every day and monitors her diet closely. Tanya makes her way in the world by taking part-time jobs and has to watch where the money goes. It is crucial that she takes the right amount of minerals and vitamins each month to stay fit and healthy. The amounts have been determined by her coach. He suggests that future Olympic champions should absorb at least 120 milligrams (mg) of vitamins and at least 880 mg of minerals each month. To make sure she follows this regime Tanya relies on two food supplements. One is in solid form and has the trade name Solido and the other is in liquid form marketed under the name Liquex. Her problem is to decide how much of each she should purchase each month to satisfy her coach.

The classic diet problem is to organize a healthy diet and pay the lowest price for it. It was a prototype for problems in linear programming, a subject developed in the 1940s that is now used in a wide range of applications.

At the beginning of March Tanya takes a trip to the supermarket and checks out Solido and Liquex. On a back of a packet of Solido she finds out it contains 2 mg vitamins and 10 mg minerals, while a carton of Liquex contains 3 mg vitamins and 50 mg minerals. She dutifully fills her trolley with 30 packets of Solido and 5 cartons of Liquex to keep herself going for the month. As she proceeds towards the checkout she wonders if she has the right amount. First she calculates how many vitamins she has in

	Solido	Liquex	Requirements
Vitamins	2 mg	3 mg	120 mg
Minerals	10 mg	50 mg	880 mg

timeline

AD1826
Fourier anticipates linear programming; Gauss solves linear equations by Gaussian elimination

1902
Farkas gives a solution of inequality systems

the trolley. In the 30 packets of Solido she has $2 \times 30 = 60$ mg vitamins and in the Liquex, $3 \times 5 = 15$. Altogether she has $2 \times 30 + 3 \times 5 = 75$ mg vitamins. Repeating the calculation for minerals, she has $10 \times 30 + 50 \times 5 = 550$ mg minerals.

As the coach required her to have at least 120 mg vitamins and 880 mg minerals, she needs more packets and cartons in the trolley. Tanya's problem is juggling the right amounts of Solido and Liquex with the vitamin and mineral requirements. She goes back to the health section of the supermarket and puts more packets and cartons into her trolley. She now has 40 packets and 15 cartons. Surely this will be OK? She recalculates and finds she has $2 \times 40 + 3 \times 15 = 125$ mg vitamins and $10 \times 40 + 50 \times 15 = 1150$ mg minerals. Now Tanya certainly satisfies her coach's recommendation and has even exceeded the required amounts.

Feasible solutions The combination (40, 15) of foods will enable Tanya to satisfy the diet. This is called a possible combination, or a 'feasible' solution. We have seen already that (30, 5) is not a feasible solution so there is a demarcation between the two types of combinations – feasible solutions in which the diet is fulfilled and non-feasible solutions in which it is not.

Tanya has many more options. She could fill her trolley with only Solido. If she did this she would need to buy at least 88 packets. The purchase (88, 0) satisfies both requirements, because this combination would contain $2 \times 88 + 3 \times 0 = 176$ mg vitamins and $10 \times 88 + 50 \times 0 = 880$ mg minerals. If she bought only Liquex she would need at least 40 cartons, the feasible solution (0, 40) satisfies both vitamin and mineral requirements, because $2 \times 0 + 3 \times 40 = 120$ mg vitamins and $10 \times 0 + 50 \times 40 = 2000$ mg minerals. We may notice that the intake of vitamins and minerals is not met *exactly* with any of these possible combinations though the coach will certainly be satisfied Tanya is having enough.

Optimum solutions Money is now brought into the situation. When Tanya gets to the checkout she must pay for the purchases. She notes that the packets and cartons are equally priced at £5 each. Of the feasible combinations we have found so far (40, 15), (88, 0) and (0, 40) the bills would be £275, £440 and £200, respectively so the best solution so far will be to buy no Solido and 40 cartons of Liquex. This will be the least cost purchase and the dietary

945

igler solves the diet problem
a heuristic method

1947

Dantzig formulates the simplex
method and solves the diet
problem by linear programming

1984

Karmarker derives a new algorithm for
solving linear programming problems

requirement will be achieved. But how much food to buy has been hit and miss. On the spur of the moment Tanya has tried various combinations of Solido and Liquex and figured out the cost in these cases only. Can she do better? Is there a possible combination of Solido and Liquex that will satisfy her coach and at the same time cost her the least? What she would like to do is to go home and analyse the problem with a pencil and paper.

Linear programming problems
Tanya's always been coached to visualize her goals. If she can apply this to winning Olympic gold, why not to mathematics? So she draws a picture of the feasible region. This is possible because she is only considering two foods. The line AD represents the combinations of Solido and Liquex that contain exactly 120 mg vitamins. The combinations above this line have more than 120 mg vitamins. The line EC represents the combinations that contain exactly 880 mg minerals. The combinations of foods that are above both these lines is the feasible region and represents all the feasible combinations Tanya could buy.

Problems with the framework of the diet problem are called linear programming problems. The word 'programming' means a procedure (its usage before it became synonymous with computers) while 'linear' refers to the use of straight lines. To solve Tanya's problem with linear programming, mathematicians have shown that all we need to do is to work out the size of the food bill at the corner points on Tanya's graph. Tanya has discovered a new feasible solution at the point B with coordinates (48, 8) which means that she could purchase 48 packets of Solido and 8 cartons of Liquex. If she did this she would satisfy her diet *exactly* because in this combination there is 120 mg of vitamins and 880 mg of minerals. At £5 for both a packet and a carton this combination would cost her £280. So the optimum purchase will remain as it was before, that is, she should purchase no Solido at all and 40 cartons of Liquex at a total cost of £200, even though she will have 1120 mg of vitamins in excess of the 880 mg required.

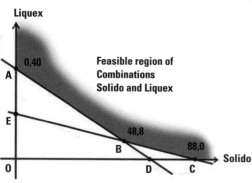

The optimum combination ultimately depends on the relative costs of the supplements. If the cost per packet of Solido went down to £2 and Liquex went up to £7 then the bills for the corner point combinations A (0, 40), B (48, 8) and C (88, 0) would be respectively £280, £152 and £176.

The best purchase for Tanya with these prices is 48 packets of Solido and 8 cartons of Liquex, with a bill of £152.

History In 1947 the American mathematician George Dantzig, then working for the US Air Force, formulated a method for solving linear programming problems called the simplex method. It was so successful that Dantzig became known in the West as the father of linear programming. In Soviet Russia, cut off during the Cold War, Leonid Kantorovich independently formulated a theory of linear programming. In 1975, Kantorovich and the Dutch mathematician Tjalling Koopmans were awarded the Nobel Prize for Economics for work on the allocation of resources, which included linear programming techniques.

Tanya only considered only two foods – two variables – but nowadays problems involving thousands of variables are commonplace. When Dantzig found his method there were few computers but there was the Mathematical Tables Project – a decade-long job creation scheme which began in New York in 1938. It took a team of some ten human calculators working for 12 days with hand calculators to solve a diet problem in nine 'vitamin' requirements and 77 variables.

While the simplex method and its variants have been phenomenally successful, other methods have also been tried. In 1984 the Indian mathematician Narendra Karmarkar derived a new algorithm of practical significance, and the Russian Leonid Khachiyan proposed one of chiefly theoretical importance.

The basic linear programming model has been applied to many situations other than choosing a diet. One type of problem, the transportation problem, concerns itself with transporting goods from factories to warehouses. It is of a special structure and has become a field in its own right. The objective in this case is to minimize the cost of transportation. In some linear programming problems the objective is to maximize (like maximizing profit). In other problems the variables only take integer values or just two values 0 or 1, but these problems are quite different and require their own solution procedures.

It remains to be seen whether Tanya Smith wins her gold medal at the Olympic Games. If so, it will be another triumph for linear programming.

The condensed idea
Keeping healthy
at least cost

46 The travelling salesperson

James Cook, based in Bismarck (North Dakota, USA), is a super salesperson for the Electra company, a manufacturer of carpet cleaners. The fact that he has won salesperson of the year for three years straight is evidence of his ability. His sales area takes in the cities of Albuquerque, Chigaco, Dallas and El Paso, and he visits each of them in a round trip once a month. The question he sets himself is how to make the trip and at the same time minimize the total mileage travelled. It is the classical travelling salesperson problem.

James has drawn up a mileage chart showing distances between the cities. For example, the distance between Bismarck and Dallas is 1020 miles, found at the intersection (shaded) of the Bismarck column with the Dallas row.

Albuquerque				
883	Bismarck			
1138	706	Chicago		
580	1020	785	Dallas	
236	1100	1256	589	El Paso

The greedy method Being a practical person, James Cook sketches a map of the sales area but doesn't worry about accuracy so long as it tells him roughly where the cities are and the distances between them. One route he often takes starts from Bismarck travelling to Chicago, Albuquerque, Dallas and El Paso in turn before returning to Bismarck. This is the route BCADEB but he realizes this trip of 4113 miles in total is expensive in terms of the distance travelled. Can he do better?

Making a plan of the sales area should not disguise the fact that James is not in the mood for detailed planning – he wants to get out there and sell. Looking at a map in his office in Bismarck he sees that the nearest city is Chicago. It is 706 miles away as against 883 to Albuquerque, 1020 miles to Dallas, and 1100

c.ᴬᴰ**1810**	**1831**	**1926**
Charles Babbage mentions the problem as an interesting one	The travelling salesperson problem appears as a practical problem	Borůvka introdu greedy algorith

miles to El Paso. He starts for Chicago immediately without an overall plan. When he gets to Chicago and completes his business there he looks for the nearest city to go to. He chooses Dallas in preference to Albuquerque and El Paso, because it is 785 miles from Chicago, and is nearer than the other options.

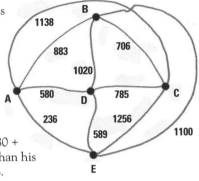

Once in Dallas he has notched up 706 + 785 miles. He then has to choose between Albuquerque and El Paso. He chooses Albuquerque because it is closer. From Albuquerque he has to go to El Paso whereupon he has visited all the cities and, job done, he returns to Bismarck. His total mileage is 706 + 785 + 580 + 236 + 1100 = 3407 miles. This route BCDAEB is much shorter than his previous route and he has made a saving on carbon emissions too.

This way of thinking is often called the greedy method of finding a short route. This is because James Cook's decision is always a local one – he is in a certain city and looks for the best route out of that city. With this method he never attempts to look forward by more than one step at a time. It is not strategic because it takes no overall account of the best route. The fact that he finished in El Paso meant he was forced to take a long route back to Bismarck. So a shorter route has been found, but is it the shortest route? James is intrigued.

James sees how he can take advantage of there being only five cities involved. With so few it is possible to list all the possible routes and to then select the shortest one. With five cities there are only 24 routes to examine or just 12 if we count a route and its reverse as equivalent. This is permissible because both have the same total mileage. The method serves James Cook well and he learns that the route BAEDCB (or its reverse BCDEAB) is actually optimum, being only 3199 miles long.

Back in Bismarck James realizes his travel is taking too long. It is not the distance he wants to save, but time. He draws up a new table that gives the travel time between the different cities in his territory.

When the problem was focused on mileages, James knew that the sum of the distances along two sides of a triangle is always greater than the

Albuquerque				
12 (road)	Bismarck			
6 (air)	2 (air)	Chicago		
2 (air)	4 (air)	3 (air)	Dallas	
4 (road)	3 (air)	5 (air)	1 (air)	El Paso

1954

Dantzig and Dijkstra propose methods for attacking the travelling salesperson problem

1971

Cook formulates the P versus NP concept for algorithms

2004

David Applegate solves the problem for all 24,978 cities in Sweden

length of the third side; in this case the graph is called Euclidean and much is known about solution methods. This is not the case when the problem is one of time. Flying on main routes is often faster than side routes and James Cook notes that in going from El Paso to Chicago it is quicker to fly via Dallas. The so-called triangle inequality does not apply.

The greedy method applied to the time-problem produces a total time of 22 hours on the route BCDEAB, whereas there are two distinct optimum routes BCADEB and BCDAEB totalling 14 hours each. Of these two routes, the first is 4113 miles, and the second 3407 miles. James Cook is happy that by choosing BCDAEB he has saved the most. As a future project he will consider the route with least cost.

From seconds to centuries The real difficulty associated with the travelling salesperson problem occurs when there is a large number of cities. Because James Cook is such a brilliant employee it is not long before he is promoted to being a supervisor. He now has to visit 13 cities from Bismarck instead of the previous 4. He is unhappy about using the greedy method, and prefers to look at a complete listing of the routes. He sets out to list the possible routes for his 13 cities. He soon discovers there would be as many as 3.1×10^9 routes to examine. Put another way, if a computer took one second to print out a route it would take about a century to print them all. A problem with 100 cities as input would tie up the computer for millennia.

Some sophisticated methods have been applied to the travelling salesperson problem. Exact methods have been given which apply to 5000 cities or less and one has even successfully dealt with a particular problem of 33,810 cities, though the computer power required in this case was colossal. Non-exact methods produce routes which are within range of the optimum with a specified probability. Methods of this type have the advantage of being able to handle problems with millions of cities.

Computational complexity Looking at the problem from a computer's point of view, let's just think about the time it might take to find a solution. Simply listing all possible routes is a worst case scenario. James has discovered that this brute force method for 13 cities would take almost a century to complete. If we threw in an extra 2 cities the time would go up to over 20,000 years!

Of course these estimates will depend on the actual computer used, but for n cities the time taken rises in line with n factorial (the number you get by multiplying together all whole numbers from 1 to n). We calculated 3.1×10^9

routes for 13 cities. Deciding whether each route is the shortest yet found becomes a problem of *factorial* time – and it will be a long time.

Other methods are available for attacking the problem in which the time for n cities rises with 2^n (2 multiplied by itself n times) so for 13 cities this would involve the order of 8192 decisions (8 times more than for 10 cities). A method with this complexity is called an *exponential* time algorithm. The holy grail of these 'combinatorial optimization problems' is to find an algorithm which depends not on the nth power of 2, but on a fixed power of n. The smaller the power the better; for example, if the algorithm varied according to n^2 then, in the case of 13 cities, this would amount to only 169 decisions – less than twice the time taken for 10 cities. A method of this 'complexity' is said to be conducted in *polynomial* time – problems solved this way are 'quick problems' and could take 3 minutes, rather than centuries.

The class of problems that can be *solved* by a computer in polynomial time is denoted by P. We don't know if the travelling salesperson problem is one of these. No one has come up with a polynomial time algorithm for it but nor have they been able to show that none exists.

A wider class denoted by NP consists of problems whose solutions can be *verified* in polynomial time. The travelling salesperson problem is definitely one of these because checking whether a *given* route is a shorter distance than any given distance can be done in polynomial time. You just add the distances along the given route and compare it with the given number. *Finding* and *verifying* are two different operations: for instance, it is easy to verify that $167 \times 241 = 40{,}247$ but finding the factors of 40,247 is a different proposition.

Is *every* problem verifiable in polynomial time able to be solved in polynomial time? If this were true the two classes P and NP would be identical and we could write $P = NP$. Whether $P = NP$ is the burning question of the day for computer scientists. More than half the profession think this is not true: they believe there are problems out there that can be checked in polynomial time but cannot be solved in polynomial time. It is such an outstanding problem that the Clay Mathematics Institute has offered a prize of $1,000,000 to prove whether $P = NP$ or $P \neq NP$.

The condensed idea
Finding the best route

47 Game theory

Some said Johnny was the smartest person alive. John von Neumann was a child prodigy who became a legend in the mathematical world. When people heard that he arrived at a meeting in a taxi having just scribbled out his 'minimax theorem' in game theory, they just nodded. It was exactly the sort of thing von Neumann did. He made contributions to quantum mechanics, logic, algebra, so why should game theory escape his eye? It didn't – with Oskar Morgenstern he coauthored the influential *Theory of Games and Economic Behavior.* In its widest sense game theory is an ancient subject, but von Neumann was key to sharpening the theory of the 'two-person zero-sum game'.

Two-person zero-sum games It sounds complicated, but a two-person zero-sum game is simply one 'played' by two people, companies, or teams, in which one side wins what the other loses. If A wins £200 then B loses that £200; that's what zero-sum means. There is no point in A cooperating with B – it is pure competition with only winners and losers. In 'win–win' language A wins £200 and B wins –£200 and the sum is 200 + (–200) = 0. This is the origin of the term 'zero-sum'.

Let's imagine two TV companies ATV and BTV are bidding to operate an extra news service in either Scotland or England. Each company must make a bid for one country only and they will base their decision on the projected increased size of their viewing audiences. Media analysts have estimated the increased audiences and both companies have access to their research. These are conveniently set down in a 'payoff table' and measured in units of a million viewers.

timeline

AD**1713**	**1944**
Waldegrave gives the first mathematical solution of a two-player game	von Neumann and Morgenstern publish *Theory of Games and Economic Behavior*

		BTV	
		Scotland	England
ATV	Scotland	+5	−3
	England	+2	+4

If both ATV and BTV decide to operate in Scotland then ATV will gain 5 million viewers, but BTV will lose 5 million viewers. The meaning of the minus sign, as in the payoff −3, is that ATV will *lose* an audience of 3 million. The + payoffs are good for ATV and the − payoffs are good for BTV.

We'll assume the companies make their *one-off* decisions on the basis of the payoff table and that they make their bids simultaneously by sealed bids. Obviously both companies act in their own best interests.

If ATV chooses Scotland the worst that could happen would be a loss of 3 million; if it bids for England, the worst would be a gain of 2 million. The obvious strategy for ATV would be to choose England (row 2). It couldn't do worse than gain 2 million viewers whatever BTV chooses. Looking at it numerically, ATV works out −3 and 2 (the row minimums) and chooses the row corresponding to the maximum of these.

BTV is in a weaker position but it can still work out a strategy that limits its potential losses and hope for a better payoff table next year. If BTV chooses Scotland (column 1) the worst that could happen would be a loss of 5 million; if it chooses England, the worst would be loss of 4 million. The safest strategy for BTV would be to choose England (column 2) for it would rather lose an audience of 4 million than 5 million. It couldn't do worse than lose 4 million viewers whatever ATV decides.

These would be the safest strategies for each player and, if followed, ATV would gain 4 million extra viewers while BTV loses them.

A beautiful mind

John F. Nash (b.1928) whose troubled life was portrayed in the 2001 movie *A Beautiful Mind* won the Nobel Prize for Economics in 1994 for his contributions to game theory.

Nash and others extended game theory to the case of more than two players and to games where cooperation between players occurs, including ganging up on a third player. The 'Nash equilibrium' (like a saddle point equilibrium) gave a much broader perspective than that set down by von Neumann, resulting in a greater understanding of economic situations.

1950
Tucker poses the prisoner's dilemma and Nash proposes the Nash equilibrium

1982
Maynard Smith publishes *Evolution and the Theory of Games*

1994
Nash is awarded the Nobel Prize for Economics for his work on game theory

When is a game determined? The following year, the two TV companies have an added option – to operate in Wales. Because circumstances have changed there is a new payoff table.

		BTV			
		Wales	Scotland	England	row minimum
ATV	Wales	+3	+2	+1	+1
	Scotland	+4	−1	0	−1
	England	−3	+5	−2	−3
	column maximum	+4	+5	+1	

As before, the safe strategy for ATV is to choose the row which maximizes the worst that can happen. The maximum from {+1, −1, −3} is to choose Wales (row 1). The safe strategy for BTV is to choose the column which minimizes from {+4, +5, +1}. That is England (column 3).

By choosing Wales, ATV can *guarantee* to win no less than 1 million viewers whatever BTV does, and by choosing England (column 3), BTV can *guarantee* to lose no more than 1 million viewers whatever ATV does. These choices therefore represent the *best* strategies for each company, and in this sense the game is determined (but it is still unfair to BTV). In this game the

$$\text{maximum of } \{+1, -1, -3\} = \text{minimum of } \{+4, +5, +1\}$$

and both sides of the equation have the common value of +1. Unlike the first game, this version has a 'saddle-point' equilibrium of +1.

Repetitive games The iconic repetitive game is the traditional game of 'paper, scissors, stone'. Unlike the TV company game which was a one-off, this game is usually played half a dozen times, or a few hundred times by competitors in the annual World Championships.

In 'paper, scissors, stone', two players show either a hand, two fingers, or a fist, each symbolizing paper, scissors or stone. They play simultaneously on the count of three: paper

	paper	scissors	stone	row minimum
paper	draw = 0	lose = -1	win = +1	−1
scissors	win = +1	draw = 0	lose = −1	−1
stone	lose = −1	win = +1	draw = 0	−1
column maximum	+1	+1	+1	

draws with paper, is defeated by scissors (since scissors can cut paper), but defeats stone (because it can wrap stone). If playing 'paper' the payoffs are therefore 0, −1, +1, which is the top row of our completed payoff table.

There is no saddle point for this game and no obvious *pure* strategy to adopt. If a player always chooses the same action, say paper, the opponent will detect this and simply choose scissors to win every time. By von Neumann's 'minimax theorem' there is a 'mixed strategy' or a way of choosing different actions based on probability.

According to the mathematics, players should choose randomly but overall the choices of paper, scissors, stone should each be made a third of the time. 'Blind' randomness may not always be the best course, however, as world champions have ways of choosing their strategy with a little 'psychological' spin. They are good at second-guessing their opponents.

When is a game *not* zero-sum? Not every game is zero-sum – each player sometimes has their own separate payoff table. A famous example is the 'prisoner's dilemma' designed by A.W. Tucker.

Two people, **A**ndrew and **B**ertie, are picked up by the police on suspicion of highway robbery and held in separate cells so they cannot confer with each other. The payoffs, in this case jail sentences, not only depend on their individual responses to police questioning but on how they *jointly* respond. If **A** confesses and **B** doesn't then **A** gets only a one year sentence (from **A**'s payoff table) but **B** is sentenced to ten years (from **B**'s payoff table). If **A** doesn't confess but **B** does, the sentences go the other way around. If both confess they get four years each but if neither confesses and they both maintain their innocence they get off scot-free!

A		B	
		confess	not confess
A	confess	+4	+1
	not confess	+10	0

B		B	
		confess	not confess
A	confess	+4	+10
	not confess	+1	0

If the prisoners could cooperate they would take the optimum course of action and not confess – this would be the 'win–win' situation.

the condensed idea
Win–win mathematics

48 Relativity

When an object moves, its motion is measured relative to other objects. If we drive along a major road at 70 miles per hour (mph) and another car is driving outside us at 70 mph in the same direction, our speed relative to this car is zero. Yet we are both travelling at 70 mph relative to the ground. And our speed is 140 mph relative to a car driving at 70 mph on the opposite carriageway. The theory of relativity changed this way of thinking.

The theory of relativity was set out by the Dutch physicist Hendrik Lorentz in the late 19th century, but the definitive advance was made by Albert Einstein in 1905. Einstein's famous paper on special relativity revolutionized the study of how objects move, reducing Newton's classical theory, a magnificent achievement in its time, to a special case.

Back to Galileo To describe the theory of relativity we take a tip from the master himself: Einstein loved to talk about railway trains and thought experiments. In our example, Jim Diamond is on board a train travelling at 60 mph. From his seat at the back of the train he walks towards the cafeteria car at 2 mph. His speed is 62 mph relative to the ground. On returning to his seat Jim's speed relative to the ground will be 58 mph because he is walking in the opposite direction to the train's travel. This is what Newton's theory tells us. Speed is a relative concept and Jim's direction of motion determines whether you add or subtract.

Because all motion is relative, we talk about a 'frame of reference' as the viewpoint from which a particular motion is measured. In the one-dimensional motion of the train moving along a straight track we can think of a fixed frame of reference positioned at a railway station and a distance x and a time t in terms of this reference frame. The zero position is determined by a point marked on the platform and the time read from the station clock. The distance/time coordinates relative to this reference frame at the station are (x, t).

timeline

C.AD 1632	1676	1687
Galileo gives the 'Galilean transformations' for falling bodies	Römer calculates the speed of light from observations of the moons of Jupiter	Newton's *Principia* describes the classical laws of motion

There is also a reference frame on board the train. If we measure distance from the end of the train and time by Jim's wristwatch there would be another set of coordinates $(\bar{x}, \bar{t})$. It is also possible to synchronize these two coordinate systems. When the train passes the mark on the platform, then $x = 0$ and the station clock is at $t = 0$. If Jim sets $\bar{x} = 0$ at this point, and puts $\bar{t} = 0$ on his wristwatch, there is now a connection between these coordinates.

As the train passes through the station Jim sets off for the cafeteria car. We can calculate how far he is from the station after five minutes. We know that the train is travelling at 1 mile per minute, so in that time it has travelled 5 miles *and* Jim has walked $\bar{x} = \frac{1}{6}$ of a mile (from his speed of 2 mph multiplied by time $\frac{5}{60}$). So in total Jim is a distance (x) which is $5\frac{1}{6}$ miles from the station. The relation between x and $\bar{x}$ is therefore given by $x = \bar{x} + v \times t$ (here $v = 60$). Turning the equation around to give the distance Jim has travelled relative to the reference frame on the train, we get

$$\bar{x} = x - v \times t$$

The concept of time in the classical Newtonian theory is a one-dimensional flow from the past to the future. It is universal for all and it is independent of space. Since it is an absolute quantity, Jim's time on board the train is the same for the station master on the platform t, so

$$\bar{t} = t$$

These two formulae for $\bar{x}$ and $\bar{t}$, first derived by Galileo, are types of equations called transformations, as they transform quantities from one reference frame to another. According to Newton's classical theory, the speed of light should be expected to obey these two Galilean transformations for $\bar{x}$ and $\bar{t}$.

By the 17th century people recognized that light had speed, and its approximate value was measured in 1676 by the Danish astronomer Ole Römer. When Albert Michelson measured the speed of light more accurately in 1881, he found it was 186,300 miles per second. More than this, he became aware that the transmission of light was very different from the transmission of sound. Michelson found that, unlike the speed of our observer on the moving train, the direction of the light beam had no bearing on the speed of light at all. This paradoxical result had to be explained.

881	**1887**	**1905**	**1915**
Michelson measures the speed of light with great accuracy	The Lorentz transformations are first written down	Einstein publishes *On the electrodynamics of moving bodies,* the paper that describes special relativity	Einstein publishes *The field equations for gravitation,* describing general relativity

$$\alpha = \frac{1}{\sqrt{1 - \frac{v^2}{c^2}}}$$

The Lorentz factor

The special theory of relativity Lorentz set out the mathematical equations which governed the connection between the distance and time when one frame of reference moves at a constant speed v relative to another. These transformations are very similar to the ones we have already worked out but involve a (Lorentz) factor depending on v and the speed of light, c.

Enter Einstein The way Einstein dealt with Michelson's findings about the speed of light was to adopt it as a postulate:

The speed of light is the same value for all observers and is independent of direction.

If Jim Diamond flicked a torch on and off while passing through the station on his speeding train, firing the light beam down the carriage in the direction the train was moving, he would measure its speed as c. Einstein's postulate says that the watching stationmaster on the platform would also measure the beam's speed as c, not as c + 60 mph. Einstein also assumed a second principle:

One frame of reference moves with constant speed in relation to another.

The brilliance of Einstein's 1905 paper was due in part to the way he approached his work, being motivated by mathematical elegance. Sound waves travel as vibrations of molecules in the medium through which the sound is being carried. Other physicists had assumed light also needed some medium to travel through. No one knew what it was, but they gave it a name – the luminiferous aether.

Einstein felt no need to assume the existence of the aether as the medium for transmitting light. Instead, he deduced the Lorentz transformations from the two simple principles of relativity and the whole theory unfolded. In particular, he showed that the energy of a particle E is determined by the equation $E = \alpha \times mc^2$. For the energy of a body at rest (when $v = 0$ and so $\alpha = 1$), this leads to the iconic equation showing that mass and energy are equivalent:

$$E = mc^2$$

Lorentz and Einstein were both proposed for the Nobel Prize in 1912. Lorentz had already been given it in 1902, but Einstein had to wait until 1921 when he was finally awarded the prize for work on the photoelectric effect (which he had also published in 1905). That was quite a year for the Swiss patent clerk.

Einstein vs Newton For observations on slow moving railway trains there is only a very small difference between Einstein's relativity theory and the classical Newtonian theory. In these situations the relative speed v is so small compared with the speed of light that the value of the Lorentz factor α is almost 1. In this case the Lorentz equations are virtually the same as the classical Galilean transformations. So for slow speeds Einstein and Newton would agree with each other. Speeds and distances have to be very large for the differences between the two theories to be apparent. Even the record breaking French TGV train has not reached these speeds yet and it will be a long time in the development of rail travel before we would have to discard the Newtonian theory in favour of Einstein's. Space travel will force us to go with Einstein.

The general theory of relativity Einstein published his general theory in 1915. This theory applies to motion when frames of reference are allowed to *accelerate* in relation to each other and links the effects of acceleration with those of gravity.

Using the general theory Einstein was able to predict such physical phenomena as the deflection of light beams by the gravitational fields of large objects such as the Sun. His theory also explained the motion of the axis of Mercury's rotation. This precession could not be fully explained by Newton's theory of gravitation and the force exerted on Mercury by the other planets. It was a problem that had bothered astronomers since the 1840s.

The appropriate frame of reference for the general theory is that of the four-dimensional space–time. Euclidean space is flat (it has zero curvature) but Einstein's four-dimensional space–time geometry (or Riemannian geometry) is curved. It displaces the Newtonian force of gravity as the explanation for objects being attracted to each other. With Einstein's general theory of relativity it is the curvature of space–time which explains this attraction. In 1915 Einstein launched another scientific revolution.

the condensed idea
The speed of light is absolute

49 Fermat's last theorem

We can add two square numbers together to make a third square. For instance, $5^2 + 12^2 = 13^2$. But can we add two cubed numbers together to make another cube? What about higher powers? Remarkably, we cannot. Fermat's last theorem says that for any four whole numbers, x, y, z and n, there are no solutions to the equation $x^n + y^n = z^n$ when n is bigger than 2. Fermat claimed he'd found a 'wonderful proof', tantalizing the generations of mathematicians that followed including a ten-year-old boy who read about this mathematical treasure hunt one day in his local library.

Fermat's last theorem is about a Diophantine equation, the kind of equation which poses the stiffest of all challenges. These equations demand that their solutions be whole numbers. They are named after Diophantus of Alexandria whose *Arithmetica* became a milestone in the theory of numbers. Pierre de Fermat was a 17th-century lawyer and government official in Toulouse in France. A versatile mathematician, he enjoyed a high reputation in the theory of numbers, and is most notably remembered for the statement of the last theorem, his final contribution to mathematics. Fermat proved it, or thought he had, and he wrote in his copy of Diophantus's *Arithmetica* 'I have discovered a truly wonderful proof, but the margin is too small to contain it.'

Fermat solved many outstanding problems, but it seems that Fermat's last theorem was not one of them. The theorem has occupied legions of mathematicians for three hundred years, and has only recently been proved. This proof could not be written in any margin and the modern techniques required to generate it throw extreme doubt on Fermat's claim.

timeline

AD1665	1753	1825	1839
Fermat dies, leaving no record of his 'wonderful proof'	Euler proves the case for $n = 3$	Legendre and Dirichlet independently prove the case for $n = 5$	Lamé prove case for $n =$

The equation $x + y = z$ How can we solve this equation in three variables x, y and z? In an equation we usually have one unknown x but here we have three. Actually this makes the equation $x + y = z$ quite easy to solve. We can choose the values of x and y any way we wish, add them to get z and these three values will give a solution. It is as simple as that.

For example if we choose $x = 3$ and $y = 7$, the values $x = 3$, $y = 7$ and $z = 10$ make a solution of the equation. We can also see that some values of x, y and z are not solutions of the equation. For example $x = 3$, $y = 7$ and $z = 9$ is not a solution because these values do not make the left-hand side $x + y$ equal the right hand side z.

The equation $x^2 + y^2 = z^2$ We'll now think about squares. The square of a number is that number multiplied by itself, the number which we write as x^2. If $x = 3$ then $x^2 = 3 \times 3 = 9$. The equation we are thinking of now is not $x + y = z$, but

$$x^2 + y^2 = z^2$$

Can we solve this as before, by choosing values for x and y and computing z? With the values $x = 3$ and $y = 7$, for example, the left-hand side of the equation is $3^2 + 7^2$ which is $9 + 49 = 58$. For this z would have to be the square root of 58 ($z = \sqrt{58}$) which is approximately 7.6158. We are certainly entitled to claim that $x = 3$, $y = 7$ and $z = \sqrt{58}$ is a solution of $x^2 + y^2 = z^2$ but unfortunately Diophantine equations are primarily concerned with whole number solutions. As $\sqrt{58}$ is not a whole number, the solution $x = 3$, $y = 7$ and $z = \sqrt{58}$ will not do.

The equation $x^2 + y^2 = z^2$ is connected with triangles. If x, y and z represent the lengths of the three sides of a right-angled triangle they satisfy this equation. Conversely, if x, y and z satisfy the equation then the angle between x and y is a right angle. Because of the connections with Pythagoras's theorem solutions for x, y and z are called Pythagorean triples.

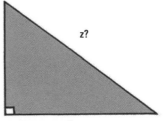

How can we find Pythagorean triples? This is where the local builder comes to the rescue. Part of the builder's equipment is the ubiquitous

843

1907

1908

1994

ummer claims he has
roved the theorem, but
Dirichlet exposes a flaw

von Lindemann
claims a proof, but is
shown to be wrong

Wolfskehl offers a prize
for solutions within the
next 100 years

Wiles finally proves
the theorem

3-4-5 triangle. The values $x = 3$, $y = 4$ and $z = 5$ turn out to be a solution of the kind we are looking for because $3^2 + 4^2 = 9 + 16 = 5^2$. From the converse, a triangle with dimensions 3, 4 and 5, must include a right angle. This is the mathematical fact that the builder uses to build his walls at right angles.

In this case we can break up a 3 × 3 square, and wrap it around a 4 × 4 square to make a 5 × 5 square.

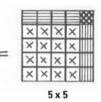

3 x 3   **4 x 4**   **5 x 5**

There are other whole number solutions $x^2 + y^2 = z^2$. For example $x = 5$, $y = 12$ and $z = 13$ is another solution because $5^2 + 12^2 = 13^2$ and in fact there are an infinite number of solutions to the equation. The builder's solution $x = 3$, $y = 4$ and $z = 5$ holds pride of place since it is the smallest solution, and is the only solution composed of consecutive whole numbers. There are many solutions where two numbers are consecutive, such as $x = 20$, $y = 21$ and $z = 29$ as well as $x = 9$, $y = 40$ and $z = 41$, but no others with all three.

From feast to famine It looks like a small step to go from $x^2 + y^2 = z^2$ to $x^3 + y^3 = z^3$. So, following the idea of reassembling one square around another to form a third square, can we pull off the same trick with a cube? Can we reassemble one cube around another to make a third? It turns out this can't be done. The equation $x^2 + y^2 = z^2$ has an infinite number of different solutions but Fermat was unable to find even one whole number example of $x^3 + y^3 = z^3$. Worse was to follow, and Leonhard Euler's lack of findings led him to phrase the last theorem:

> *There is no solution in whole numbers to the equation $x^n + y^n = z^n$ for all values of n higher than 2.*

One way to approach the problem of proving this is to start on the low values of n and move forward. This was the way Fermat went to work. The case $n = 4$ is actually simpler than $n = 3$ and it is likely Fermat had a proof in this case. In the 18th and 19th centuries, Euler filled in the case $n = 3$, Adrien-Marie Legendre completed the case $n = 5$ and Gabriel Lamé proved the case $n = 7$. Lamé initially thought he had a proof of the general theorem but was unfortunately mistaken.

Ernst Kummer was a major contributor and in 1843 submitted a manuscript claiming he had proved the theorem in general, but Dirichlet pointed out a gap in the argument. The French Academy of Sciences offered a prize of 3000 francs for a valid proof, eventually awarding it to Kummer for his worthy attempt. Kummer proved the theorem for all primes less than 100 (and other values) but excluding the irregular primes 37, 59 and 67. For example, he could not prove there were no whole numbers which satisfied $x^{67} + y^{67} = z^{67}$. His failure to prove the theorem generally opened up valuable techniques in abstract algebra. This was perhaps a greater contribution to mathematics than settling the question itself.

Ferdinand von Lindemann, who did prove that the circle could not be squared (see page 22) claimed to have proved the theorem in 1907 but was found to be in error. In 1908 Paul Wolfskehl bequeathed a 100,000 marks prize to be awarded to the first provider of a proof, a prize made available for 100 years. Over the years, something like 5000 proofs have been submitted, checked, and all returned to the hopefuls as being false.

The proof While the link with Pythagoras's theorem only applies for $n = 2$, the link with geometry proved the key to its eventual proof. The connection was made with the theory of curves and a conjecture put forward by two Japanese mathematicians Yutaka Taniyama and Goro Shimura. In 1993 Andrew Wiles gave a lecture on this theory at Cambridge and included his proof of Fermat's theorem. Unhappily this proof was wrong.

The similarly named French mathematician André Weil dismissed such attempts. He likened proving the theorem with climbing Everest and added that if a man falls short by 100 yards he has not climbed Everest. The pressure was on. Wiles cut himself off and worked on the problem incessantly. Many thought Wiles would join that throng of the nearly people.

With the help of colleagues, however, Wiles was able to excise the error and substitute a correct argument. This time he convinced the experts and proved the theorem. His proof was published in 1995 and he claimed the Wolfskehl prize just inside the qualifying period to become a mathematical celebrity. The ten-year-old boy sitting in a Cambridge public library reading about the problem years before had come a long way.

the condensed idea
Proving a marginal point

50 The Riemann hypothesis

The Riemann hypothesis represents one of the stiffest challenges in pure mathematics. The Poincaré conjecture and Fermat's last theorem have been conquered but not the Riemann hypothesis. Once decided, one way or the other, elusive questions about the distribution of prime numbers will be settled and a range of new questions will be opened up for mathematicians to ponder.

The story starts with the addition of fractions of the kind

$$1 + \frac{1}{2} + \frac{1}{3}$$

The answer is 1⅚ (approximately 1.83). But what happens if we keep adding smaller and smaller fractions, say up to ten of them?

$$1 + \frac{1}{2} + \frac{1}{3} + \frac{1}{4} + \frac{1}{5} + \frac{1}{6} + \frac{1}{7} + \frac{1}{8} + \frac{1}{9} + \frac{1}{10}$$

Using only a handheld calculator, these fractions add up to approximately 2.9 in decimals. A table shows how the total grows as more and more terms are added.

Number of terms	Total (approximate)
1	1
10	2.9
100	5.2
1,000	7.5
10,000	9.8
100,000	12.1
1,000,000	14.4
1,000,000,000	21.3

The series of numbers

$$1 + \frac{1}{2} + \frac{1}{3} + \frac{1}{4} + \frac{1}{5} + \frac{1}{6} + \dots$$

is called the harmonic series. The harmonic label originates with the Pythagoreans who believed that a musical string divided by a half, a third, a quarter, gave the musical notes essential for harmony.

timeline

AD **1854**
Riemann begins his work on the zeta function

1859
Riemann proves key solutions lie in a critical strip and puts forward his conjecture

1896
De la Vallée-Poussin and Hadamard show all important zeros lie *within* Riemann's critical strip

In the harmonic series, smaller and smaller fractions are being added but what happens to the total? Does it grow beyond all numbers, or is there a barrier somewhere, a limit that it never rises above? To answer this, the trick is to group the terms, doubling the runs as we go. If we add the first 8 terms (recognizing that $8 = 2 \times 2 \times 2 = 2^3$) for example

$$S_{2^3} = 1 + \frac{1}{2} + \left(\frac{1}{3} + \frac{1}{4}\right) + \left(\frac{1}{5} + \frac{1}{6} + \frac{1}{7} + \frac{1}{8}\right)$$

(where S stands for sum) and, because ⅓ is bigger than ¼ and ⅕ is bigger than ⅛ (and so on), this is greater than

$$1 + \frac{1}{2} + \left(\frac{1}{4} + \frac{1}{4}\right) + \left(\frac{1}{8} + \frac{1}{8} + \frac{1}{8} + \frac{1}{8}\right) = 1 + \frac{1}{2} + \frac{1}{2} + \frac{1}{2}$$

So we can say

$$S_{2^3} > 1 + \frac{3}{2}$$

and more generally

$$S_{2^k} > 1 + \frac{k}{2}$$

If we take $k = 20$, so that $n = 2^{20} = 1,048,576$ (more than a million terms), the sum of the series will only have exceeded 11 (see table). It is increasing in an excruciatingly slow way – but, a value of k can be chosen to make the series total beyond *any* preassigned number, however large. The series is said to diverge to infinity. By contrast, this does not happen with the series of squared terms

$$1 + \frac{1}{2^2} + \frac{1}{3^2} + \frac{1}{4^2} + \frac{1}{5^2} + \frac{1}{6^2} + \dots$$

We are still using the same process: adding smaller and smaller numbers together, but this time a limit is reached, and this limit is less than 2. Quite dramatically the series converges to $\pi^2/6 = 1.64493\dots$

In this last series the power of the terms is 2. In the harmonic series the power of the denominators is silently equal to 1 and this value is critical. If the power increases by a minuscule amount to a number just above 1 the series converges, but if the power decreases by a minuscule amount to a value just below 1, the

1900	1914	2004
Hilbert places the hypothesis in his list of key problems for mathematicians to solve	Hardy proves there are infinitely many solutions along Riemann's line	The first 10 trillion zeros are verified to be on the critical line

series diverges. The harmonic series sits on the boundary between convergence and divergence.

The Riemann zeta function

The celebrated Riemann zeta function $\zeta(s)$ was actually known to Euler in the 18th century but Bernhard Riemann recognized its full importance. The ζ is the Greek letter zeta, while the function is written as:

$$\zeta(s) = 1 + \frac{1}{2^s} + \frac{1}{3^s} + \frac{1}{4^s} + \frac{1}{5^s} + \cdots$$

Various values of the zeta function have been computed, most prominently, $\zeta(1) = \infty$ because $\zeta(1)$ is the harmonic series. The value of $\zeta(2)$ is $\pi^2/6$, the result discovered by Euler. It has been shown that the values of $\zeta(s)$ all involve π when s is an even number while the theory of $\zeta(s)$ for odd values of s is far more difficult. Roger Apéry proved the important result that $\zeta(3)$ is an irrational number but his method did not extend to $\zeta(5)$, $\zeta(7)$, $\zeta(9)$, and so on.

The Riemann hypothesis

The variable s in the Riemann zeta function represents a real variable but this can be extended to represent a complex number (see page 32). This enables the powerful techniques of complex analysis to be applied to it.

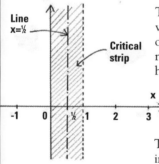

Line
x=½

Critical
strip

-1   0   ½ 1   2   3

x

The Riemann zeta function has an infinity of zeros, that is, an infinity of values of s for which $\zeta(s) = 0$. In a paper presented to the Berlin Academy of Sciences in 1859, Riemann showed all the important zeros were complex numbers that lay in the critical strip bounded by $x = 0$ and $x = 1$. He also made his famous hypothesis:

All the zeros of the Riemann zeta function $\zeta(s)$ lie on the line $x = ½$; the line along the middle of the critical strip.

The first real step towards settling this hypothesis was made in 1896 independently by Charles de la Vallée-Poussin and Jacques Hadamard. They showed that the zeros must lie on the interior of the strip (so x could not equal 0 or 1). In 1914, the English mathematician G.H. Hardy proved that an infinity of zeros lie along the line $x = ½$ though this does not prevent there being an infinity of zeros lying off it.

As far as numerical results go, the non-trivial zeros calculated by 1986 (1,500,000,000 of them) do lie on the line $x = ½$ while up-to-date calculations have verified this is also true for the first 100 billion zeros. While these experimental results suggest that the conjecture is reasonable, there is still the possibility that it may be false. The conjecture is that *all* zeros lie on this critical line, but this awaits proof or disproof.

Why is the Riemann hypothesis important?
There is an unexpected connection between the Riemann zeta function $\zeta(s)$ and the theory of prime numbers (see page 36). The prime numbers are 2, 3, 5, 7, 11 and so on, the numbers only divisible by 1 and themselves. Using primes, we can form the expression

$$\left(1 - \frac{1}{2^s}\right) \times \left(1 - \frac{1}{3^s}\right) \times \left(1 - \frac{1}{5^s}\right) \times \dots$$

and this turns out to be another way of writing $\zeta(s)$, the Riemann zeta function. This tells us that knowledge of the Riemann zeta function will throw light on the distribution of prime numbers and enhance our understanding of the basic building blocks of mathematics.

In 1900, David Hilbert set out his famous 23 problems for mathematicians to solve. He said of his eighth problem, 'if I were to awaken after having slept for five hundred years, my first question would be: Has the Riemann hypothesis been proven?'

Hardy used the Riemann hypothesis as insurance when crossing the North sea after his summer visit to his friend Harald Bohr in Denmark. Before leaving port he would send his friend a postcard with the claim that he had just proved the Riemann hypothesis. It was a clever each way bet. If the boat sank he would have the posthumous honour of solving the great problem. On the other hand, if God did exist he would not let an atheist like Hardy have that honour and would therefore prevent the boat from sinking.

The person who can rigorously resolve the issue will win a prize of a million dollars offered by the Clay Mathematics Institute. But money is not the driving force – most mathematicians would settle for achieving the result and a very high position in the pantheon of great mathematicians.

the condensed idea
The ultimate challenge

Glossary

Algebra Dealing with letters instead of numbers so as to extend arithmetic, algebra is now a general method applicable to all mathematics and its applications. The word 'algebra' derives from '*al-jabr*' used in an Arabic text of the ninth century AD.

Algorithm A mathematical recipe; a set routine for solving a problem.

Argand diagram A visual method for displaying the two-dimensional plane of complex numbers.

Axiom A statement, for which no justification is sought, that is used to define a system. The term 'postulate' served the same purpose for the Greeks but for them it was a self-evident truth.

Base The basis of a number system. The Babylonians based their number system on 60, while the modern base is 10 (decimal).

Binary number system A number system based on *two* symbols, 0 and 1, fundamental for computer calculation.

Cardinality The number of objects in a set. The cardinality of the set {a, b, c, d, e} is 5, but cardinality can also be given meaning in the case of infinite sets.

Chaos theory The theory of dynamical systems that appear random but have underlying regularity.

Commutative Multiplication in algebra is commutative if $a \times b = b \times a$, as in ordinary arithmetic (e.g. $2 \times 3 = 3 \times 2$). In many branches of modern algebra this is not the case (e.g. matrix algebra).

Conic section The collective name for the classical family of curves which includes circles, straight lines, ellipses, parabolas and hyperbolas. Each of these curves is found as cross-sections of a cone.

Corollary A minor consequence of a theorem.

Counterexample A single example that disproves a statement. The statement 'All swans are white' is shown to be false by producing a black swan as a counterexample.

Denominator The bottom part of a fraction. In the fraction $\frac{3}{7}$, the number 7 is the denominator.

Differentiation A basic operation in Calculus which produces the derivative or rate of change. For an expression describing how distance depends on time, for example, the derivative represents the velocity. The derivative of the expression for velocity represents acceleration.

Diophantine equation An equation in which solutions have to be whole numbers or perhaps fractions. Named after the Greek mathematician Diophantus of Alexandria (c.AD 250).

Discrete A term used in opposition to 'continuous'. There are gaps between discrete values, such as the gaps between the whole numbers 1, 2, 3, 4, . . .

Distribution The range of probabilities of events that occur in an experiment or situation. For example, the Poisson distribution gives the probabilities of x occurrences of a rare event happening for each value of x.

Divisor A whole number that divides into another whole number exactly. The number 2 is a divisor of 6 because $6 \div 2 = 3$. So 3 is another because $6 \div 3 = 2$.

Empty set The set with no objects in it. Traditionally denoted by ϕ, it is a useful concept in set theory.

Exponent A notation used in arithmetic. Multiplying a number by itself, 5×5 is written 5^2 with the exponent 2. The expression $5 \times 5 \times 5$ is written 5^3, and so on. The notation may be extended: for example, the number $5^{\frac{1}{2}}$ means the square root of 5. Equivalent terms are power and index.

Fraction A whole number divided by another, for example $\frac{3}{4}$.

Geometry Dealing with the properties of lines, shapes, and spaces, the subject was formalized in Euclid's *Elements* in the third century BC. Geometry pervades all of mathematics and has now lost its restricted historical meaning.

Greatest common divisor, *gcd* The *gcd* of two numbers is the largest number which divides into both exactly. For example, 6 is the *gcd* of the two numbers 18 and 84.

Hexadecimal system A number system of base 16 based on 16 symbols, 0, 1, 2, 3, 4, 5, 6, 7, 8, 9, A, B, C, D, E, and F. It is widely used in computing.

Hypothesis A tentative statement awaiting either proof or disproof. It has the same mathematical status as a conjecture.

Imaginary numbers Numbers involving the 'imaginary' $i = \sqrt{-1}$. They help form the complex numbers when combined with ordinary (or 'real') numbers.

Integration A basic operation in Calculus that measures area. It can be shown to be the inverse operation of differentiation.

Irrational numbers Numbers which cannot be expressed as a fraction (e.g. the square root of 2).

Iteration Starting off with a value a and repeating an operation is called iteration. For example, starting with 3 and repeatedly adding 5 we have the iterated sequence 3, 8, 13, 18, 23, . . .

Lemma A statement proved as a bridge towards proving a major theorem.

Matrix An array of numbers or symbols arranged in a square or rectangle. The arrays can be added together and multiplied and they form an algebraic system.

Numerator The top part of a fraction. In the fraction ⅗, the number 3 is the numerator.

One-to-one correspondence The nature of the relationship when each object in one set corresponds to exactly one object in another set, and vice versa.

Optimum solution Many problems require the best or optimum solution. This may be a solution that minimizes cost or maximizes profit, as occurs in linear programming.

Place-value system The magnitude of a number depends on the position of its digits. In 73, the place value of 7 means '7 tens' and of 3 means '3 units'.

Polyhedron A solid shape with many faces. For example, a tetrahedron has four triangular faces and a cube has six square faces.

Prime number A whole number that has only itself and 1 as divisors. For example, 7 is a prime number but 6 is not (because $6 \div 2 = 3$). It is customary to begin the prime number sequence with 2.

Pythagoras's theorem If the sides of a right-angled triangle have lengths x, y and z then $x^2 + y^2 = z^2$ where z is the length of the longest side (the hypotenuse) opposite the right angle.

Quaternions Four-dimensional imaginary numbers discovered by W.R. Hamilton.

Rational numbers Numbers that are either whole numbers or fractions.

Remainder If one whole number is divided by another whole number, the number left over is the remainder. The number 17 divided by 3 gives 5 with remainder 2.

Sequence A row (possibly infinite) of numbers or symbols.

Series A row (possibly infinite) of numbers or symbols added together.

Set A collection of objects: for example, the set of some items of furniture could be F = {chair, table, sofa, stool, cupboard}.

Square number The result of multiplying a whole number by itself. The number 9 is a square number because $9 = 3 \times 3$. The square numbers are 1, 4, 9, 16, 25, 36, 49, 64, . . .

Square root The number which, when multiplied by itself, equals a given number. For example, 3 is the square root of 9 because $3 \times 3 = 9$.

Squaring the circle The problem of constructing a square with the same area as that of a given circle – using only a ruler for drawing straight lines and a pair of compasses for drawing circles. It cannot be done.

Symmetry The regularity of a shape. If a shape can be rotated so that it fills its original imprint it is said to have rotational symmetry. A figure has mirror symmetry if its reflection fits its original imprint.

Theorem A term reserved for an established fact of some consequence.

Transcendental number A number that cannot be the solution of an algebraic equation, like $ax^2 + bx + c = 0$ or one where x has an even higher power. The number π is a transcendental number.

Twin primes Two prime numbers separated by at most one number. For example, the twins 11 and 13. It is not known whether there is an infinity of these twins.

Unit fraction Fractions with the top (numerator) equal to 1. The ancient Egyptians partly based their number system on unit fractions.

Venn diagram A pictorial method (balloon diagram) used in set theory.

x–y axes The idea due to René Descartes of plotting points having an x-coordinate (horizontal axis) and y-coordinate (vertical axis).

Index

Bold page numbers indicate a glossary entry

This 2013 edition is published for Barnes & Noble, Inc.
by Quercus Editions Ltd.

Picture credits: p.101: iStockphotos

Edited by Keith Mansfield
Designed by Patrick Nugent
Illustrations and diagrams by Tony Crilly and Patrick Nugent
Proofread by Anna Faherty
Index by Ingrid Lock

ISBN 978-1-4351-4739-3

Manufactured in China

2 4 6 8 10 9 7 5 3 1